MUSCLECAR
COLOR · HISTORY

BOSS & COBRA JET MUSTANGS
302, 351, 428, and 429

Dr. John Craft

MBI Publishing Company

For Gordon Bender; gone but not forgotten

First published in 1996 by MBI Publishing Company, 729 Prospect Avenue, PO Box 1, Osceola, WI 54020-0001 USA

© Dr. John Craft, 1996

All rights reserved. With the exception of quoting brief passages for the purposes of review no part of this publication may be reproduced without prior written permission from the Publisher.

The information in this book is true and complete to the best of our knowledge. All recommendations are made without any guarantee on the part of the author or Publisher, who also disclaim any liability incurred in connection with the use of this data or specific details.

We recognize that some words, model names and designations, for example, mentioned herein are the property of the trademark holder. We use them for identification purposes only. This is not an official publication.

MBI Publishing Company books are also available at discounts in bulk quantity for industrial or sales-promotional use. For details write to Special Sales Manager at Motorbooks International Wholesalers & Distributors, 729 Prospect Avenue, PO Box 1, Osceola, WI 54020-0001 USA.

Library of Congress Cataloging-in-Publication Data
Craft, John Albert.
  Boss & Cobra Jet Mustangs: 302, 351, 428, and 429/ John Craft.
     p. cm. —(MBI Publishing Company muscle car color history series)
  Includes index.
  ISBN 0-7603-0050-X (pbk.: alk. paper)
  1. Mustang automobile—History. I. Title. II. Series.
TL215.M8C725   1996
629.222'2—dc20   96-717

On the front cover: A 1970 Boss 302, a car conceived, designed, and built to tackle the Trans-Am circuit. This car is owned by David Samuels.
On the frontispiece: One of the trickest options one could add to the Cobra Jet engine was the shaker hood, which bobbed and shuddered with each powerful pulse of the big-block engine underneath.
On the title page: In May of 1969, Parnelli Jones (15) and Mark Donahue (6) do battle at Michigan International, where the Boss 302 debuted. Jones and the fledgling Boss won the race, but Donahue and his Camaro won the Championship. *Craft Collection*
On the back cover: Above: Another look at Parnelli Jones in a Boss 302, this one his 1970 car. Jones and his Boss went on to win the 1970 Trans-Am Championship. *Craft Collection Below:* The mighty 428 Cobra Jet engine, complete with shaker hood.

Printed in China

Contents

	Acknowledgments	6
One	Before the Thunder	9
Two	The 1969 Boss 302	21
Three	The 1970 Boss 302	55
Four	The Boss 429	77
Five	The Boss 351	95
Six	The 428 Cobra Jet Mustang	103
Seven	The 429 Cobra Jet Mustang	121
Appendix A	Engine Specifications	124
Appendix B	CJ/SCJ Parts Comparison	126
Appendix C	Production Figures	127

Acknowledgments

This book was made possible in large measure by the support and encouragement of Jim Smart over the past 15 years. Back in the 1980s, Jim was the Editor of *Mustang Monthly* magazine, then the premier publication devoted to the preservation and restoration of first generation Mustangs. It was to Jim that I made my first submissions as a writer. For reasons not entirely clear to me even today, Jim chose to publish those articles rather than summarily consigning them to the nearest trash bin. As compared to some of lesser stature in the Mustang world who have abandoned their one time focus on first generation pony cars in the fickle pursuit of ad copy, the publications that Jim continues to oversee have remained steadfast in their dedication to 1965–73 Mustangs. His current title, Petersen Publication's *Mustang & Fords*, is today the only periodical still devoted exclusively to first generation Mustangs and high performance Fords of the Total Performance era. It is in that same vein that this book was written.

Lee Iacocca hoped to attract youthful car buyers in 1964 with the sports car reputation he created for the new Mustang line. The Mustang II styling exercise embodied many of the "sporty car" cues that Iacocca hoped would win the hearts (and pocket books) of sports car enthusiasts. Jim Smart Collection

ONE

Before the Thunder

Ford's sporty Mustang was conceived during the informal meetings held by Lee Iacocca's "Fairlane Group" in the bar of the Fairlane Inn, just down the street from Ford world headquarters in Dearborn. The purpose of those regular eight-member meetings was to chart the Ford Motor Company's course in the face of changing market forces and consumer demographics anticipated during the 1960s.

Iacocca was convinced that Ford was in desperate need of a sporty, youth-oriented car designed to capture the fancy (and pocketbooks) of the then-burgeoning 15- to 29-year-old age block. He announced that fact at a 1961 meeting of the Fairlane Group. Research had revealed the post-war baby boom generation would be reaching car buying age by the beginning of the sixties and that by 1970 the number of sub-30-year-old car buyers would swell by 40 percent. Iacocca foresaw a distinct business opportunity in those predictions. He and fellow Fairlane Group members soon set the corporation on a path that led directly to the creation of the Ford Mustang.

The T-5 project was a direct outgrowth of the Fairlane Group's weekly meetings. As originally configured, the lithe little two-seat T-5 roadster was designed to provide true sports car performance. Built around a tubular steel frame that sported four-wheel independent suspension, an integral roll bar, and an amidships-mounted V-4 engine, the T-5—later called the Mustang I—was quite unlike anything Ford had ever built. Unfortunately for the string back driving glove and meerschaum pipe group, the Mustang I was too different for Ford to use as anything more than a design study and conversation starter on the show car circuit.

Though the regular production Mustangs that began rolling off of UAW assembly lines across the country two years after the Mustang I made its public debut at Watkins Glen in 1962 were nominally related to the little roadster, that was about the extent of their similarity. Fact of the matter was, in RPO form, Ford's new for 1965 Mustangs were essentially little more than tarted up Falcons. Indeed, most of a Mustang's unit production chassis and solid rear axle/unequal length A-frame underpinnings came directly from the Falcon parts bin with precious few (if any) modifications. The same could be said of the sleepy straight six cylinder and marginally more aroused 260ci V-8s that powered the very first Mustang road cars off of the assembly line. In short, though marketed as sporting high performance vehicles, the very first Mustangs had very little to boast about in sports car circles. And that was a matter of no small concern to Mr. Iacocca and his Fairlane Group visionaries.

Winning the hearts and minds of the baby boom generation would take considerably more than the sensible but tepid motoring offered up by the Falcon line. The solution was obvious. Since the new Mustang line wasn't "born" with racing bona fides, it would have to earn respect with track victories.

Iacocca reasoned that the brisk early sixties sales of imported sports cars like the Jaguar, MG, Triumph and Austin-Healy (and even Chevrolet's eccentric Corvair) indicated a healthy interest of the buying public in the SCCA-style road racing. So he early on decided that the new Mustang line needed to be "blooded" in sports car competition. Unfortunately for Iacocca, there were few people at Ford who had the ability to develop and oversee a Mustang racing program. That was largely due to Ford division chief Robert McNamara's 1957 decision to pack up Ford's racing tent and withdraw from all forms of factory backed motorsports. Despite the fact that McNamara left the corporation to work in the Kennedy and Johnson administrations as Secretary of Defense, Ford's dominance in NASCAR and other forms of racing in the 1950s was but a bittersweet memory in 1964.

Shortly after replacing McNamara as Ford GM in 1960, Iacocca had granted an audience to a brash Texas sports car racer named Shelby who had finished first at LeMans in 1959. The subject of that first meeting was power for a hybrid, Anglo-American sports car that Shelby was working on in California. Though Iacocca initially agreed only to provide a handful of Ford's new 260ci small block engines, by 1964 Shelby's band of West Coast hot rodders had become Ford's de facto sports car racing arm. Shelby American's racing operation was so successful in international endurance racing that Enzo Ferrari's blood-red GT cars very nearly lost the World Sports Car championship to Shelby's blue and white striped Cobra roadsters in 1964 (truth be known, Ferrari would have lost the title that year were it not for the factory's behind-the-scenes maneuvering with the sanctioning body).

It was only natural, then, for Iacocca to turn to Shelby early in 1964 to seek his help with the soon-to-be-released Mustang line. By December of 1964 the race shop at Shelby American had built a sufficient number of modified early 1965 2+2s to satisfy the SCCA's homologation rules. Road race victory number one for the Shelby modified GT350s (so called because it was approximately 350ft from Shelby's office to the converted airport hanger Shelby Mustang assembly line at LAX) came on Valentine's Day 1965. It was the first of many and, for the next

Winning sports car bona fides for Mr. Iacocca's Mustang was first delegated to Carroll Shelby. He delivered with a series of SCCA B/Production driving championships garnered by his GT350R Mustangs.

three years, Shelby Mustang drivers dominated the SCCA's B/Production racing class taking three consecutive championships.

Those Shelby GT350 road race wins coupled with a handful of highly touted victories in European rally competition, did much to achieve the racing bona fides that Iacocca had lusted after for the new Mustang line. Their cumulative effect was to link an air of sportiness with even the most mundane of six-cylinder Mustang coupes and untold thousands more of America's first "pony car" were sold as a result.

Within days of Ford's unveiling of the Mustang line at the 1964 New York World's Fair, Plymouth countered with the introduction of its very own "pony car," the Barracuda. Those two models—coupled with the pre-existing Ford Falcon, Dodge Dart, Plymouth Valiant and Chevrolet Corvair lines—were seen by SCCA officials as the basis for an American production-based sedan racing series. The SCCA called the new competition category the Trans-American Sedan Championship and subdivided it into two different categories: under two liters of engine displacement (Group I) and over two liters (Group II). As you might have guessed, Ford Mustangs with the 260ci (and ultimately 289ci) engines were classified as over two liter cars along with their other domestic class rivals.

The all-new racing classification perfectly suited Lee Iacocca's desire to establish the Mustang as a sports car of note. Better yet, the other cars in the over two-liter class were the Mustang's Big Three competitors. What better way to prove the new pony car's sporting mettle than by trouncing the contenders on the track?

Jerry Titus added many trophies to his mantelpiece in his Shelby team 1967 Trans-Am notchback. He also secured Ford's second straight Trans-Am title with the car. Photo Courtesy of SAAC

The very first Trans-Am race was held on the fabled airport road course in Sebring, Florida, in March of 1966. Though no factory sponsored Mustangs were on the starting grid that day, racing dentist Dr. Dick Thompson's number 20, early 1965 Mustang was the fastest car during qualifying. During the race, A.J. Foyt's notchback Mustang led the first 70mi of the event. Unfortunately for independent Mustang drivers like Thompson, Chrysler cars—particularly the number 44 Dodge Dart that Bob Tullius drove to victory that day—proved to be more than a match for their home built Trans-Am ponies early in the season.

The lack of initial success persuaded Iacocca to put the same type of factory backed support into the new Trans-Am (T/A) series that had been so successful in the B/Production (B/P) ranks. Once again, Carroll Shelby was called upon to prepare a fleet of specially modified notchbacks for use in Trans-Am competition. To that end, 17 notchbacks (one prototype and 16 production) were diverted from the San Jose assembly line and shipped to Shelby's airport complex. Once there, the cars received the by now race-proven modifications that had led to the GT350s first B/P title in 1965. When a sufficient number of the cars had been converted to racing specifications, the SCCA homologated them for competition in May of 1966. The converted notchbacks were then offered up to the general public for sale at $6,414 a copy (it's interesting to note that a race ready R-Model GT350 actually stickered for less than a Group II Trans-Am-prepared Mustang at $5,950!).

Once the cars were at the track, Shelby team cars were chauffeured by drivers like Jerry Titus, Bob

For a time during the 1967 Trans-Am season it appeared that Ford teams would be forced to surrender their title to Bud Moore's factory-backed Cougar team. Ford made sure that didn't happen a second time by pulling the funding from Moore's competition cats the following season. Craft Collection

Johnson, and Don Pike. Other Ford-backed notchbacks were campaigned at selected events by NASCAR stars like David Pearson, Curtis Turner, and Wendell Scott. The end result was Ford's first Trans-Am manufacturer's title, which was won in dramatic fashion. Going into the final race at the now-defunct Riverside International Raceway (RIR), the points race for the manufacturer's title was a dead heat. With the sales-generating bragging rights to the Trans-Am series championship on the line, Iacocca sent in the full Shelby team to RIR to ensure a favorable outcome. When the green flag fell, 34 Trans-Am cars headed off towards turn one. The Group II contingent consisted of 14 entries, eight of them Mustangs. Jerry Titus led the three-car Shelby American contingent. Titus, a part-time racer and full-time editor of *Sports Car Graphic* magazine sat on the pole of the race with a 91.85mph qualifying lap around the nine-turn RIR road course. The 1:49.9sec it took Titus to execute that circuit in his specially prepared Shelby team notchback was nearly two full seconds faster than any other car in the field, which made him a natural favorite to take the checkered flag first. Titus made it harder than it should have been by flooding the car's carburetor during the LeMans-style start LeMans starts begin with a foot race to the car. The driver then fires up the engine, and takes off. By the time Titus got the flooded engine started and the car underway, he had fallen far behind the race leaders. Nonetheless, Titus overcame the poor start and a mid-race broken oil filter to finish first. When the dust settled, the final points standings placed Ford ahead of the Chrysler Plymouth team by a convincing seven points in the coveted manufacturer's championship.

The following year saw the Mustang line's first body redesign. Longer, lower, and wider were the styling watchwords of the day and that's pretty much how the new for 1967 Mustang line looked. Purists weren't entirely pleased, but the changes were mostly positive for racing purposes. The increased tire track translated into greater cornering resolve out on the track, for example. The new Mustang's big-block capable engine bay also permitted better cooling and

Parnelli Jones, Ed Leslie, and Dan Gurney drove red and silver Trans-Am Cougars for Bud Moore during the 1967 SCCA season and gave Ford's Mustang teams a run for their money. Craft Collection

work access when fitted with an SCCA Trans-Am spec small block engine.

Shelby American was again assigned the task of building a fleet of Group II legal notchbacks for use on the increasingly popular Trans-Am circuit. As in 1966, Group II Trans-Am Mustangs were required to retain both their basic stock silhouettes and back seats (this latter requirement being primarily responsible for the absence of GT350s in the Trans-Am series, by the way). Stock appearance notwithstanding, all manner of mechanical modifications were either made possible or mandatory by the rules book. Engine size was still limited to five liters but significant changes were permitted, including the use of a free-breathing twin four-barrel induction system. With a set of headers and full-race crank, rods, and pistons working in concert with the dual carburetors, horsepower figures flirted with the 400 mark. All told, Shelby American produced 26 1967 Group II Mustangs. As in 1966, most of the coupes produced were sold to well-heeled racers in the general public. Four of the cars were retained by Shelby American and served as SAAC competition models for the 1967 motorsports season.

Increasing interest in the Trans-Am series led the sanctioning body to increase the number of races on the circuit to 12. Along with the expanded schedule came a new scoring system that awarded the championship to manufacturers rather than individual drivers in the hopes that Detroit's Big Three would become increasingly involved in sponsorship of the series. It worked. Ford was eager to score more victories on the circuit and upped the competitive ante by calling famed NASCAR team owner, Bud Moore, away from the Grand National ranks to oversee a team of all new Cougars for drivers Dan Gurney, Rufus "Parnelli" Jones, and Ed Leslie. In addition to the extra in-house Lincoln and Mercury competition, Mustang Group II drivers also faced off against Chevrolet drivers campaigning that manufacturer's new-for-1967 Camaros. In that number was a young engineer named Mark Donohue who early in the season fielded his home-built Camaro. By the end of 1967, Donohue had teamed up with another Chevrolet racer named Roger Penske to campaign a squad of lightning fast blue and yellow Camaros.

Jerry Titus gave up his "day job" at *Sports Car Graphic* to assume full-time driving duties for the Shelby American team, although he kept his typewriter

Shelby American coupes continued to be the chassis of choice for Ford's factory cars on the 1968 Trans-Am circuit. The cars were powered by the new Tunnel Port 302s. Daytona Speedway Archives

handy for the weekly Trans-Am series reports he'd been contracted to write for *Autoweek* magazine. Titus opened the season for Ford and Shelby with a fourth at Daytona where the Bud Moore-prepped Cougars had made a strong initial showing. The rivalry between the two Fomoco factions was fierce and overshadowed both the class win scored by Bob Tullius' Dart and the efforts of the new Chevrolet teams. In March, Titus won at Sebring and by season's end had scored enough points to win a second straight Trans-Am title for Ford. Second place in the points race came the red and silver Cougars campaigned by Gurney, Jones, and Leslie. The final SCCA tally placed Group II Mustangs just two points ahead of Moore's team Cougars which is why, perhaps, Moore was sent back to the NASCAR ranks for 1968 and the Cougar team retired. Chevrolet teams finished a distant third in the championship, but that was soon to change—primarily due to a change in powerplants made by Ford's racing division during the off-season.

A New Engine for 1968

By 1968 the Trans-American Sedan series evolved from a secondary series to one of the most highly-coveted championships. The Shelby team dropped the pretext of signing "independent drivers" and openly campaigned a factory-backed team under Shelby Racing Company colors. As in 1967, Jerry Titus got the nod as the number one driver. Dan Gurney, Parnelli Jones, Peter Revson, and a number of other road racers also spent time at the helm of Shelby Trans-Am cars in 1968. The slight styling changes that distinguished the 1967 and 1968 Mustang lines made it possible for the team to retain (and simply "remodel") some of the Group II coupes that had seen service the year before. In addition to those chassis, five new notchbacks were built for duty on the 1968 circuit. All the cars campaigned by the Shelby team that year featured the wider wheels and fender bulges permitted by the revised rules book and they all tipped the scales at a new 2,800 lb. minimum weight.

Though only superficially changed from their 1967 form in terms of cosmetics, Shelby's 1968 Trans-Am coupes were radically different from their predecessors in mechanical configuration. From the first moment that Mustang Trans-Am cars (or B/Production cars for that matter) had graced a racing grid, they were motivated by high performance versions of the

Ford's first and greatest success during the 1968 season came at Daytona where Jerry Titus finished fourth overall behind a troika of long-tailed Porsches. The rest of the season was characterized by a series of disastrous engine failures that allowed Mark Donohue to break Ford's lock on the Trans-Am title. Daytona Speedway Archives

289ci wedge-headed "Windsor" engine. As Ford fans will recall, those racing Hi-Po 289s featured In-line valves, fairly conventional kidney-shaped combustion chambers and equally spaced, rectangular intake and exhaust passages. Though those 289s had proven to be both reliable and powerful in competition trim, Ford engineers decided for 1968 to introduce an all new, 302ci racing small block engine that in many ways owed more to the NASCAR world than it did SCCA racing. The new power plants featured an extra 1/8in of stroke and beefed-up cylinder blocks that sported beefy four-bolt mains. Cast with the core shift closely controlled for greater cylinder thickness, the special "C-8" blocks also incorporated O-ring lands around each deck surface cylinder and water passage openings that were designed to totally eliminate the need for conventional head gaskets. A forged steel crank shaft and beefy "SK" rods were used to round out the short block along with a set of eight forged aluminum pistons.

The really big news mechanically for 1968 came bolted to the all new short block in the form of two cavernously ported "Tunnel Port" cylinder heads. Fashioned after the 427 heads used by Fomoco stock car racers in 1967 to counter the Chrysler Hemi engine, the new head castings featured radically redesigned intake passages that were so large that intake valve push-rods "tunneled" right through them in small sealed tubes. The idea was to straighten out and enlarge each intake passage in the hopes of improving high rpm performance. The revised intake passages also permitted the use of larger intake valves which also improved flow at maximum rpms. As during the 1967 Trans-Am season, two four-barrel carburetors mounted on a high riser style intake manifold were used to mix up the hydrocarbons needed for internal combustion. When properly outfitted with a set of race-tuned tubular headers and a high volt ignition system, a new Tunnel Port 302 engine was capable of cranking out nearly 450hp.

Ford racers approached the new Trans-Am season with the confident belief their new Tunnel Port engines and race-proven coupe chassis would be the class of every racing grid they graced. And, indeed, that's exactly the way the season started at the annual 24-hour-long Daytona Continental. Rick Titus quickly served notice that his red Shelby American Racing Company-backed Tunnel Port coupe was the car to

Interestingly, Ford's first Trans-Am warriors weren't factory-backed race cars. Instead, independent racers like Dr. Dick Thompson (who drove this 1964 1/2 notchback in the very first Trans-Am Sedan race at Sebring in 1966) drove road race Mustangs they'd prepared themselves.

beat in the over two-liter Group II Trans-Am class. After the clock spun twice and Titus racked up 629 laps, Titus was first in class and fourth overall behind a trio of long-tailed Porsches. Titus clung to the 907s tenaciously in the race's closing moments and responded to one of the Carrera's attempts to run him off the track by caving in one of the Porsche's alloy doors. Finish line photos show Titus and his little red Tunnel Port physically hot on the tails—but nearly 50 laps behind—the trio of 907s at the checkered flag.

Unfortunately, that first place finish turned out to be the high point of the 1968 season for Titus, the Shelby team, and Ford. Shortly after Daytona, problems with the all new Tunnel Port engine began to crop up. Oiling system deficiencies were blamed for the new engine's habit of grenading. At one point during the season, the Shelby team took to describing race weekends by the number of engines that had blown up. Six-engine weekends were commonplace and the team saw as many as seven engines self-destruct in a weekend. Team manager Lew Spencer commented at the time that eight engine weekends were out of the question only because the team was physically unable to change that many blown power plants in a two-day period. Legendary mechanic Smokey Yunick—who performed dyno work on some of the team Tunnel Ports—joked that the engines produced so little low rpm torque that you could stall the engine by grabbing the harmonic damper at idle.

The Tunnel Port's mechanical shortcomings coupled with stiff competition from factory-backed Camaro, Firebird, and Javelin teams resulted in Ford's losing the second Trans-Am championship. By the time the last drop of oil had seeped out of the Shelby team's wounded Tunnel Port engines, Mark Donohue and Roger Penske sat atop the points standings. Their team Camaros had helped score a series-winning 105 points to Ford's second place total of just 52. It was a stinging defeat and one that folks in Ford's Dearborn "Glass House" headquarters did not take lightly. After dominating just about all forms of motorsports they had entered since 1964, Ford Mustang teams had been beaten, and that was a wrong that had to be righted at all costs. By dedicating the corporation to defeating its motorsports rivals in the upcoming Trans-Am season, Ford executives laid the groundwork for the Boss 302.

Shelby American continued to build and campaign Trans-Am notchbacks in 1968. Those cars were powered by Ford's not quite legal Tunnel Port 302 motors. Though tasting early success at the grueling 24-hour Daytona Continental, the cars turned in only a mixed performance for the balance of the season.

Tunnel Port 302

Although the SCCA hoped to attract factory participation when it formed the Trans-American sedan racing series in 1965, it's doubtful that the sanctioning body imagined just how much interest the Big Three manufacturers would be paying to the Trans-Am series by 1968. Though the first cars to cut corners on the Trans-Am road racing circuit had been mostly home-built affairs that deviated only slightly from street legal trim, by the third season of the series' existence auto makers were spending truckloads of money on race-only hardware in their quest for Trans-Am victories. Acid-dipped bodies became a common (not to mention illegal) practice and all manner of specialty suspension hardware began to show up under race spec cars. Ford upped the ante even farther in 1968 by introducing an all-new racing engine built specifically for Trans-Am service. That engine was called the Tunnel Port 302 and it represented Ford's willingness to pull out all the stops in its pursuit of racing laurels.

Though similar in appearance to the 260 and 289ci "Windsor" small-block engines, the new-for-1968 Tunnel Port 302 was actually quite different from its predecessors. The cylinder block, for example, featured beefier four-bolt mains. The engine was also quite different at the top of the block. In place of head gaskets, Ford engineers opted for a gasketless "dry deck" system that used machined ring lands and O-ring gaskets to seal critical block-to-head openings. Foundry workers also paid particular attention to casting box core shift on each and every "C8FE-6010" Tunnel Port block in an effort to ensure structural strength. Once fully machined, a Tunnel Port 302 block was fitted with a forged steel crank that featured both an extra 1/8th inch of stroke (over that of a 289) and 180 degree cross-drilled hollow journals (sealed by machined plugs) that were intended to provide better lower-end lubrication. A special, significantly larger than stock damper was used to control harmonics. Forged high-dome pistons helped produce an 11:1 compression ratio and were fitted to beefed-up cast connecting rods.

Lubrication was provided by a high pressure (90–100psi) pump and an increased capacity, baffled road race pan built by Aviaid. A track specific, solid lifter camshaft and true roller timing chain system rounded out the short block's internals.

As suggested by the engine's name, the most interesting aspect of the new race engine was its cylinder heads. Like the 427

Here's a side-by-side comparison of the Tunnel Port and 351 Cleveland heads. Tunnel Port 302 engines picked up that moniker because of the way their pushrods "tunneled" through their relocated and radically enlarged intake ports. Ford created the Boss 302 engine by topping off the Tunnel Port block with a set of oval ported, 351 Cleveland head castings for 1969. As things turned out, the production based 351 heads proved to be far more potent than the exotic (and decidedly non-production) Tunnel Port castings.

Tunnel Port 302 blocks featured beefy four bolt bottom ends and carried C8FE casting numbers, just like the Boss 302 blocks that Ford began casting up one year later. Fact of the matter is, the blocks were identical. In fact, the very first Boss 302s came factory equipped with leftover 1968 C8 Tunnel Port blocks.

18

Tunnel Port big block engine used by NASCAR racers in 1967, the 302 Tunnel Port's cylinder heads incorporated radically enlarged and straightened intake runners. As a result, the intake pushrods for each cylinder had to be housed in special tubes that ran smack dab through the middle of each intake port—hence the engine's "Tunnel Port" moniker. In contrast to the radically revamped intake tract, the exhaust tract differed little from Windsor engines. The exhaust valves were slightly larger (2.15in intake and 1.57in exhaust) and were actuated by an adjustable rocker assembly similar to the system on the big-block "FE" engine.

In race trim, a Tunnel Port 302 was fed by a duo of 540cfm Holley carburetors bolted to a cast alloy, eight-venturi high rise intake manifold. When fitted with a set of free flowing headers and sparked by a mechanical advance dual-point distributor, a Tunnel Port 302 dynoed in at just short of 500hp. That was a formidable figure for a small-block engine in 1968. Fran Hernandez and Hank Lennox, two Ford engineers associated with the Tunnel Port program, were confident the new race engine would be up to the task of winning Ford's third straight Trans-Am title. Unfortunately engine failure plagued Ford Trans-Am teams throughout the season. Period reports of those failures often placed the blame for them on the engine's oiling system. Whatever the problem, it seems to have been rectified, as the highly successful Boss 302 Trans-Am cars used the same basic four-bolt short block as the Tunnel Port 302.

Race-ready Tunnel Port 302 engines like this one could be counted on for some serious top-end power. Trouble was, that was about the only place they produced horsepower. Sustained high rpm operation—then as now—more often that not results in mechanical failure of one sort or another. and that's pretty much the story of Ford team fortunes in 1968.

TWO

The 1969 Boss 302

Though the failure of Ford racers to translate the Tunnel Port 302's performance into a third straight Trans-Am title was a disappointment, better days on the Trans-Am circuit were just around the corner. The seeds for that success were laid during the frustrating 1968 model year. Those seeds were planted by a man who, up until February 6, 1968, had taken great delight in frustrating the Ford Motor Company both on and off the track. The man's name was Semon "Bunkie" Knudsen, and prior to 1968 he'd been a major figure in General Motors' motorsports and styling efforts. In the early sixties, Knudsen served as the General Manager of GM's Pontiac division when its "Tin Indians" were the fastest cars on the NASCAR circuit and spirited, youth-oriented street performers. In the late 1960s, Knudsen ascended to the Executive Vice-Presidency of General Motors. Even at that lofty corporate level, Knudsen remained an inveterate racing enthusiast. As a result, he lent his support to like-minded stylists and engineers within company even though GM was officially out of racing at the time. Chevrolet's new for 1967 Camaro was partly a result of Knudsen's enthusiasm for high performance and motorsports. As was the semi-clandestine aid that Chevrolet racers like Smokey Yunick received via the Bow Tie division's back door policy. Knudsen's enthusiasm for fun, youth-oriented automobiles also created an atmosphere for stylists like Larry Shinoda to conjure up

These styling studio photos were America's first look at a Boss 302 Mustang. Although this particular Detroit Steel Tubing-built prototype featured the, er, unusual combination of vermilion interior and bright yellow exterior, it still struck a chord with the buying public. Craft Collection

high-performance packages like the Z-28.

Chances are that Knudsen would have rounded out his manufacturing career, as his father had before him, working for GM. Fortunately for Ford race fans everywhere, Knudsen lost out to Ed Cole in a bloody power struggle at GM. Shortly thereafter Henry Ford II extended an offer to Knudsen that was too irresistible to decline. In February of 1968, Knudsen assumed responsibility for Ford's future—both on and off the track. Not surprisingly, when Knudsen set up shop in Dearborn, he brought his commitment to factory backed motorsports and exciting, youth-oriented automobiles. Knudsen also brought a number of his former GM associates. The transfer of engineers, stylists and racers began almost before Knudsen had time to try out his new Ford office. Legendary mechanic Smokey Yunick was a Knudsen contact that was put to work, as Ford experimental and high performance engine components were shipped south to his Daytona Beach-based "Best Damn Garage in Town." Knudsen also raided GM's styling studios. One of the more successful forays netted Larry Shinoda, creator of the original Corvette Stingray, Corvair Monza, and Camaro Z-28. Once at Ford, Shinoda was put in charge of Ford's Special Design Center where he was assisted by the talented quartet of designers Bill Shannon, Dick Petit, Harvey Winn and Ken Dowd. Working in concert with Special Vehicles engineers like Chuck Mountain, Ed Hall, and Bill Holbrook, Shinoda and company immediately set to work translating Knudsen's enthusiasm for high performance first into clay and then sheet metal.

Knudsen's move to Ford signaled a significant change in direction for the corporation. As Knudsen was settling into the President's suite in Dearborn, he was brimming with ideas, most of them performance-related, that would vex his former GM bosses.

He wanted cars that were invincible on the track and street going versions that looked just as unbeatable on "Woodward Avenues" all across the country. Knudsen wanted cars that were lower and wider than those currently in Ford's inventory and he was especially big on designs that featured fast back roof lines.

One of the first Ford car lines to receive Knudsen's attention was the Mustang. While Ford's first pony car had been an astonishing showroom success in 1965 and 1966, sales had steadily declined in the years that followed. Though overall styling for the 1969 Mustang line was fairly well set by the time of Knudsen's ascendancy in early 1968, he was still convinced that the line needed major changes. One of his first moves was to assign Larry Shinoda the task of revamping the Mustang line. Knudsen's hopes for the Mustang line involved much more than a mere cosmetic makeover. Knudsen told his subordinates that he wanted them to create nothing less than the best-handling street car that had ever rolled off of a UAW assembly line. That task was assigned to Howard Freers and his crew of chassis development engineers. Simultaneous to this styling and chassis work, Ford engine and foundry engineers were working on a modified version of the 1968 season's Tunnel Port 302 engine for use both on the track and in the engine bays of regular production Mustangs. Buoyed by Knudsen's support and encouragement, those three groups of Ford engineers and stylists conspired to produce what many have called the best Mustang built during the muscle car era: the Boss 302.

Styling

The basic shape of the rebodied 1969 Mustang line was set in stone well before Larry Shinoda first set pen to paper in a Ford styling studio. The theme

This styling studio shot of the Dearborn Steel Tubing built prototype captures many of the elements that showed up in the final Boss 302 package—and some that didn't. A modified version of the rear deck spoiler was part of the Boss program and so were the car's blacked out back panel, "C" stripe package, and rear window slats. The dual quad exhaust system and Kelsey Hayes mags, however, didn't make the cut. Craft Collection

followed by Shinoda's predecessors at Ford was the longer, lower, wider evolution that was typical of the major car makers of the day. The Ford Mustang was introduced as a nimble and fairly compact package, by 1969 it had grown in just about every dimension. In point of fact, one school of thought within the Ford styling ranks had the Mustang line becoming something of a down-sized Thunderbird. Which is to say, a car line long on luxury and increasingly short on performance. A Mustang station wagon was even contemplated, and one cannot imagine a farther divergence from Lee Iacocca's sports car aspirations.

Fortunately, those planned departures from Iacocca's original pony car performance theme were never taken. And Knudsen's arrival guaranteed that they would not be reconsidered in the future. The vision that Larry Shinoda brought to Ford from GM in many ways served to cement the decision to preserve the Mustang's role as Ford's in-house sports car (Carroll Shelby's Anglo-American Cobras notwithstanding).

In setting out to create a styling package for the road race-oriented Mustang that Knudsen had ordained, Larry Shinoda was influenced by both his own racing background and an interest in aerodynamics. Previous racing Mustangs (the Group II cars of 1966, 1967, and 1968) had been based on the lightweight coupe platform. Shinoda, influenced by his interest in air management, opted to base his styling package on the fastback 2+2 or sportsroof body.

His first task was to clean up the basic sheet metal package that he'd inherited from previous designers. The first order of business was the elimination of the base Mustang's curious sail panel-mounted running horse badges. Of even greater interest to Shinoda was the elimination of the simulated air intakes that regular production sportsroof Mustangs were

Larry Shinoda wasn't able to sell his Ford bosses on the "Boss" name the first time out of the box. In fact, final styling was pretty far along before that name came to be associated with Ford's new Trans-Am inspired pony car. Evidence of that fact can be found in the "Boss-less" side stripes on this styling studio prototype car. Craft Collection

The Boss 302 chassis was so fast that Ford R&D engineers insisted on wearing full race rigs (helmets and fire suits) while putting prototype cars to the test at Ford's Dearborn proving grounds. Craft Collection

slated to carry on each rear quarter panel. Reasoning that true sports car enthusiasts would not respond favorably to anything other than real performance accoutrements, Shinoda made sealing up the fake scoops a top priority and Knudsen backed him even though retooling costs for a limited production variant were high.

Next Shinoda ventured into what was then only a partially understood area of automotive performance, aerodynamics. Spoilers and the aerodynamic effects of downforce were only recent arrivals on the motorsports scene in 1967. The very first rear deck spoilers had appeared in high-speed NASCAR circles in 1966. Jim Hall's Chaparral prototype race cars and open wheeled Indy/Formula One racers had begun to experiment with such aero add-ons only shortly before their NASCAR counterparts. So Shinoda was more than a little progressive when he decided to outfit his road race-flavored Mustang with both front and rear spoilers. Ford execs initially balked at Shinoda's spoilers, insisting that the duck-tailed rear deck lip that was integral to the 1969 sportsroof design was all the aerodynamic aid the new car needed. But Shinoda was not deterred and ultimately prevailed.

Ford's Lincoln Mercury division also got its own version of the Boss 302. It was called the Cougar Eliminator. Like the Boss, the sporty new Cougar had also received the benefit of Larry Shinoda's styling genius. In addition to 302-powered Cougar Eliminators, Kar Kraft ultimately built two Boss 429-powered cats. Craft Collection

Boss cylinder heads featured poly-angle valves and generous ports. Their semi-hemispherical design helped suppress detonation.

When Ford engineers were developing the Boss engine package they explored a number of potential induction systems. One of the most radical of those was the dual four-barrel "Cross Boss" intake manifold. Quite similar to the dual-four V set up used by Chevrolet on racing Z-28 (and a few street) engines, the Cross Boss was cast to carry a pair of conventionally flanged Holley fuel mixers. Use of the Cross Boss intake required the fabrication of a special timing chain cover that moved the distributor forward far enough to clear the carburetors. A special cam extension was also part of the complicated package. Only a handful of prototype castings were made of the Cross Boss intake, but its name lived on in the form of a special intake Ford engineers cast up to carry a radical new In-line carburetor one year later.

Semon Bunkie Knudsen (left) believed that factory-backed motorsports was the key to sales floor success. He brought that belief with him (and a number of like minded folks such as Henry Smokey Yunick, on right) when he came to Ford in 1968. Good things began to happen in the Mustang line shortly thereafter. One of them was the Boss 302. Daytona Racing Archives

In final form Shinoda's trunk-mounted spoiler perched just above the body line on twin cast stanchions. Shaped in cross section like an inverted aircraft wing and fitted with drooping, canard end caps, the wing was intended to produce positive downforce when the car was at speed. Shinoda also penned a flexible plastic aero-aid for the car's front roll pan that jutted snow plow-like from beneath the bumper. Though different in configuration from the rear wing, the front spoiler was also designed to defeat destabilizing aerodynamic lift and keep the car firmly planted on the pavement at high speed. Though the jury might have been hung as to just how much work Shinoda's wings performed at road speeds, in the court of public opinion they were a smashing success that fairly shouted high performance.

Shinoda also conjured up a set of "shades" or louvers for the sportsroof body's nearly horizontal rear window that were nothing short of breathtaking. Shinoda called them sports slats and they more than a little resembled the rear window treatment on super exotic Lamborghini Muiras. Shinoda's inspiration came from the 1962 Monza Spider show car he'd penned that also featured a louvered rear window. As with his proposed wings, Shinoda's sports slats met more than a little resistance from the Fomoco bureaucracy. But once again, he persisted in his advocacy for the new rear window treatment and (with Knudsen's at least

Ford delegated development of the Boss 302 racing chassis to its "in-house" subcontractor, Kar Kraft Engineering in Brighton Michigan. Three complete Trans-Am chassis were built at Kar Kraft's 10611 Haggerty Road shop (including this beautiful black and gold car that was raced by Smokey Yunick). The suspension setting perfected by the Kar Kraft engineers ultimately showed up on race spec Boss 302s. Craft Collection

implicit backing) ultimately carried the day.

Another aspect of Shinoda's new sportsroof package was the liberal use of black semi-gloss paint both front and rear. With a nod towards the Shelby Team Group II Trans-Am racers that had gone into battle with hoods blacked out from fender to fender, Shinoda's crew cooked up a blacked-out hood panel that covered just about all of the bonnet. The semi-gloss gun wasn't put away until the new grille panel and the outboard headlight buckets had also gotten the black-out treatment. Moving aft, the top of the deck lid got a coat of low-luster black, as did the concave rear valance panel.

The total effect of Shinoda's aero add-ons and racy black-out treatment was stunning. Even so, the racy new package wasn't complete until Shinoda and his design studio crew had applied a dramatic set of signature side stripes. Inspired by the sweeping full body "C" stripes that had graced Ford's all conquering GT40 MkIV endurance racers in 1967, the Mustang stripe package flowed from the rear along the lower body line until they reached the front fenders where they widened and arched upwards before swerving back at the beltline into a tapered terminus. When coupled with a flashy set of Magnum (or in the case of some prototypes, Kelsey-Hayes) rims and a quartet of the new 60-series rubber that was just coming into vogue, Shinoda's design looked ready for the race

The first race spec Boss 302 built by Kar Kraft for duty in the 1969 season started life as an "R" code (serial number 9F02R112073) on the River Rouge assembly line. That car became Shelby American's main team car. That car is pictured here as it appeared pre-season at Riverside International Raceway on the West Coast. Notice the way its front sheet metal droops toward the pavement at a decidedly non-stock angle—all part of Kar Kraft's aerodynamic plan. Curiously, Ford racers opted to race without the rear deck spoiler that had also been homologated by street going Boss 302s. Craft Collection

course. And, of course, that's just the image that Mr. Knudsen had been looking for.

The only thing missing was a name for the car. Early candidates in the name game included the not so catchy Sports Racing Sedan and Sedan Racing Group 2 (SR-2 for short). Shinoda was appalled. For a while the name Trans-Am received serious consideration until word arrived that Pontiac's own new SCCA-inspired pony car had already co-opted that name and secured rights to it with a licensing arrangement with the sanctioning body. Luckily, Shinoda was more in tune with the market than the folks who suggested the Sports Racing Sedan moniker. As a result, he was aware of the then current California use of the word Boss to describe something that was cool, in fashion, and up to date. Shinoda wanted a name that would give his styling package a special identity; a name that would connect with the youthful market segment and convey the car's high-performance purpose. The name Boss

A race spec Boss 302's engine bay was a crowded affair crammed full of Ram Air induction bonnet, alloy radiator, auxiliary oil cooler, chassis bracing, and a dual Dominator-equipped 450hp Boss 302 engine. Craft Collection

struck him as achieving up all of those goals in four short easy to pronounce letters. And so he began to politic for its acceptance in Ford management circles. Not surprisingly, the name met the same type of stodgy resistance that his spoilers and sports slats had confronted. Getting his contemporaries to understand what the Boss name was all about took a fair amount of explaining. Eventually, Ford execs came to understand what "Boss" meant and share Shinoda's enthusiasm for the name. A short time later, Shinoda had the now famous Boss 302 name cut into the "C" stripes that had already been approved for the prototype.

Suspension

While Shinoda and his band of stylists were working on the Boss 302's cosmetics, other Ford engineers were designing the race-oriented suspension package that would help to earn the car's stripes both on and off the track. The first suspension work took place in August 1968 by Ford's specialty builder, Kar Kraft Engineering in Michigan. The Kar Kraft plant was concurrently working on what came to be called the Boss 429 (but more on that particular project a bit later). Shortly after the initial prototypes of both Boss Mustang packages were presented to Ford management for approval, development responsibility for the Boss 302's suspension was transferred from Kar Kraft to Ford's Light Vehicle Powertrain Development Section. Howard Freers was the Chief Light Car Engineer at the time and his orders from Ford President Knudsen were succinct and to the point: "build absolutely the best-handling street car available on the American market." Freers delegated that significant task to his chief Ride and Handling Engineer, Matt Donner. It was a wise decision.

Donner was much more of a hands-on racer than his engineering title might have suggested. He'd overseen suspension development for both the Mustang and Cougar lines and he'd provided input to the

Shelby team cars raced in Corporate Blue livery fetchingly accented with white Boss 302 "C" stripes. This is Shelby team car number one as it looked just before the season began in 1969. Note the National Council of Mustang Clubs decal (the race team's ostensible sponsor) on the car's front fender. Craft Collection

29

When dressed and ready for work, a race spec Boss 302 engine was an intimidating sight. Topped by more than 2,000cfm worth of Holley Dominator and fitted with a deep sump pan and tube headers, the little engine cranked out more than 450 high rpm horsepower. Craft Collection

Kar Kraft extensively modified the race Bosses suspension and chassis with more durable components, additional bracing, and widened fender wells. Unwanted rear axle movement was checked by a panhard rod assembly that pivoted on a special cross member installed on the chassis just aft of the differential. Note the lightweight aluminum air scoops used to direct cooling air to the rear discs.

racing teams that campaigned both Blue Oval pony cars in preceding SCCA seasons.

Interestingly, Boss 302 suspension development began on a 1968 GT coupe body outfitted with a small-block engine cooked up to replicate the Boss 302 engine package then under development.

Of primary concern to Donner was the Mustang line's inherent understeer. Like most American cars of the era, the Mustang chassis had been designed to produce understeer (the front tires slide rather than turn the car) at the limit. In contrast, an oversteering car tends to slide the rear end of the car at the limit and generally responds more quickly to steering input. Understeer was seen by most chassis engineers of the day as a good thing. It was widely felt that the average American driver was better able to cope with an understeering car than with a chassis setup to oversteer. That being said, understeer was definitely not the hot setup for either road course work or spirited street driving. So Donner was confronted with something of a dilemma. His task was to design the best-handling car ever made available to the driving public, but at the same time he was reluctant to create a chassis that was so twitchy and high-strung it could endanger the lives of those who tried to tame it.

Trial-and-error work involving shock compression, spring rates, sway bar diameter, and steering geometry produced a chassis combination that was capable of pulling over 1G of lateral acceleration on Ford's Dearborn test track. The Boss 302 chassis combination was so fast that extra safety precautions became necessary for test driving. Freer recalls, for example, that Donner insisted on using full road racing driver safety equipment (driving suit, helmet, full harness, etc.) because the Boss 302 was the fastest thing he'd ever driven around the Dearborn track.

In addition to the revised and significantly stiffened dampening and rebound rates that Donner dialed into the Mustang platform, he also designed a sturdier front spindle that could withstand the Boss' increased cornering loads. The combination of the improved suspension and the new F60x15 Goodyear tires also put increased stress on the chassis, so much so that the upper control arms would pull free of the shock towers under load. Donner cured the problem with a structural bracing system of metal plates that wrapped around the lower shock towers. It's interesting to note that those Boss-derived plates became standard equipment on all big-block-equipped high-performance Mustangs (Boss 429, Cobra Jet 428). In order to provide enough clearance for the wide, 60-series rubber, the front fenders incorporated re-rolled upper wheel opening arches.

Other bits and pieces of the Boss chassis package included a beefier front sway bar and, in a first for a Ford vehicle, the installation of a rear anti-sway bar. When tooling problems prevented the installation of the new rear bar for 1969 (it did show up in 1970), front sway bar diameter was decreased slightly (from $^{15}/_{16}"$ to $^{3}/_{4}"$) to compensate.

Surprisingly, the brake package fitted to the chassis was identical to the disc drum arrangement found under more mundane iterations of the breed in 1969. That fact led some road testers to complain about Boss car brake fade under rough use. Niggling aside, Donner's work was generally met with rave reviews. Perfecting the Boss chassis package ultimately used up the original 1968 mule and ten to twelve assembly line sportsroof bodies that had been destined to become Mach 1s before being pulled from the River Rouge assembly line. By late 1968, the basic Boss suspension package had been perfected and was ready to be unleashed on unsuspecting Z-28 owners and the motoring press. But not before being fitted with an all new cant-valve 302 engine.

Engine

Ford had encountered more than reliability problems with its Tunnel Port 302 engines during the 1968 season. While that high-revving engine's inability to stay in one piece had been a cause of frustration for Ford racers, its mere presence at the track had been a sore point for the sanctioning body and the balance of the Trans-Am field as a whole. You see, although the over two liter ranks of the Trans-American Sedan series was supposed to be based on regular production, American-built sedans, Ford never quite got around to building even a single street-going Tunnel Port 302 Mustang. And there's more

Smokey Yunick's Trans-Am Boss as it appears today after restoration by Brook Mossgrove—beautiful and bad to the bone. The car currently belongs to Ross Myers.

than a little doubt that the "over the counter" engine was ever actually available. Though no official sanctions were forthcoming during the 1968 season (due probably to the engine's abysmal showing at the track), things would have to be different in 1969. A thousand times different, in fact; Ford had to build and make available 1,000 Boss 302s in order for the SCCA to deem the engine "homologated" for racing purposes.

Faced with the necessity of having to build a race engine in large numbers under assembly line conditions, Ford engine and foundry engineers quickly abandoned the idea of perfecting the peaky and unreliable Tunnel Port 302's woes. But they didn't abandon that engine altogether. By fortunate coincidence, the all-new 351 Cleveland (so called because the engine was cast and assembled in that Ohio city) engine featured bore spacing and head bolt hole location that was identical to the Windsor small-block family the Tunnel Port had sprung from. Though water jacket routing was different, the engineers working on the Boss project quickly discovered that a set of poly-angle valved 351C heads would bolt onto a Tunnel Port 302 block with a minimum of modification. The planned regular production status of the 351C engine (which came on line late in 1970) coupled with the existing supply of four-bolt main journaled T/P 302 blocks seemed to be the solution to the new SCCA homologation requirement. Whether the pairing of Cleveland heads with Tunnel Port block would be successful remained a question.

In addition to their regular production status, the new Cleveland heads also offered several potential advantages over the Tunnel Port casting. The port sizes on the new Cleveland heads were actually larger than the cavernous passages of the Tunnel Port design. Those larger ports translated into equally large valve sizes. The 2.25in intake and 1.73in exhaust valves in the Cleveland castings were the largest offered on any Ford engine (big or small block) to date. Of even greater importance was the fact that those flow controllers were fitted to the heads in cant-valved fashion. The result was optimal port placement within the heads and a detonation suppressing, semi-hemi-shaped combustion chamber. In sum, the new heads seemed to be superior in just about every way to the in-line valved Tunnel Port castings. When the hybrid engine's design superiority translated into increased output on the dyno and at the track, Ford knew they had their Boss 302 engine.

R&D work with the new combination of the Tunnel Port block and 351 Cleveland heads resulted in two potent versions of the Boss 302 engine; street and race. Though both shared the same basic casting components, they were really quite different just below the surface—but not different enough to arouse the ire of the sanctioning body.

The interior of a Kar Kraft Boss 302 was race car Spartan. A single form-fitting bucket seat kept the driver firmly positioned behind the foam-padded steering wheel and close by the Ford four-speed shifter. The narrowed dash (for roll cage clearance) was gutted and housed a brace of analog gauges.

Race-spec Boss 302s were built around the same four-bolt main "C8" block that had seen service in Tunnel Port 302 trim in 1968. A forged steel crank was once again part of the Trans-Am program. Race duty Boss motors used beefy "SK" rods on each of that crank's journals and these rods were significantly stronger (and wider, in terms of "big end" bearing area) than street components. Domed forged alloy pistons were used to create compression ratios in the 12:1 range and a "Teton grind" DOZX solid lifter camshaft was mounted mid-block to oversee timing events. An Aviaid oil pan was used in concert with a set of race-spec head castings to button up the short block.

Race Boss heads for 1969 carried lightweight 2.25in intake and 1.73in exhaust valves and battleship strength valve springs. Screw-in studs captured stamped steel guide plates and provided the mounting points for 1.73 ratioed rocker arms. Special roller bearing fulcrums served as leverage points for those valve openers and lubricating fluids were directed with oiling fingers mounted to the inside of each steel valve cover.

The high rpm use envisioned for race Boss engines mandated the use of a high capacity induction system so Ford engineers went out of their way to make sure their Trans-Am 302 would be well fed. A Holley flanged alloy dual four intake was initially used for R&D work. Incredibly, Boss racers in 1969 found their heavy right feet connected to a duo of 1050cfm Holley Dominator carburetors. A special offset distributor had to be developed just to clear the front-most monster carburetor. A twin-snorkeled air box was mounted over the whole induction setup and connected via flexible tubing to the radiator support for a Ram Air effect. Free flowing stainless steel tube headers were fitted to accept the spent fires of combustion and they, in turn, handed off those carbohydrons to deliciously unmuffled dump tubes. In peak tune, a race Boss engine was good for more than 450hp and could be relied on for 8,000+ rpm back stretch blasts.

The race Boss breathed through twin Holley Dominators in 1969. Flexible tubing connected channeled cooler intake air from the front of the car into a fiberglass bonnet (not shown) mounted on top of the monstrous carbs.

As you might suspect, street Boss engines were more civilized than their road course stable mates. Yet there were still more than a few similarities between the two versions. Both engines were initially built around the beefy four-bolt main "C8" block, although the street versions would ultimately shift production to "C9" or "DO" castings. Cross-drilled forged steel cranks were used in 1969 (cross-drilling was dropped for 1970 street Bosses). Forged steel rods were fitted to the crank with beefy 3/8in bolts and worked in concert with domed forged pistons to produce a 10.5:1 compression ratio. As with racing Boss engines, a solid lifter camshaft was used.

Street Boss head castings initially carried both "C9ZE" head castings and 2.25in/1.73in sized valves (1970 Boss motors came with slightly downsized, 2.19in intakes for better low rpm response). Screw-in studs, pushrod side plates, stamped steel rocker arms, and rev-cup mounted springs were used to dress out the heads and chromed steel valve covers kept dust off of those components when the engine was in use.

Though down on total cfm in comparison to race duty engines, a street Boss' engine still came factory equipped with the largest carburetor ever fitted to a Ford high performance engine (including the Boss 429!). That particular 780cfm Holley fuel mixer was equipped with a manual choke and bolted to a dual-plane, high-rise alloy intake. Ram Air was not part of the picture for street Bosses in 1969 (shaker hoods came on the RPO scene one year later) but a vacuum-operated flapper valve was fitted to the side of the air cleaner base to permit extra air at full throttle openings. Free-flowing cast iron manifolds helped channel that extra air back towards the ozone layer via a true dual exhaust system.

In recognition of their high rpm and hard cornering ability, street Boss 302 engines came factory equipped with crank-hugging windage trays, baffled oil pans and—perhaps most importantly—a factory installed rev limiter designed to disrupt the dual point distributor's function a few hundred rpms shy of meltdown. It was not a popular item with the street racing

The first race of the 1969 season was held at Michigan in May of 1969. Ford teams went to that same track in April for a final test session to make sure they were ready to do battle. Here Bud Moore (right) looks over one of his still "C stripe-less" team cars during that test session. Craft Collection

crowd and, much to the dismay of corporate warranty bean counters, the limiter was easily and quickly disconnected. Unfortunately, revving the engine past the factory-set limit usually caused multiple valve-to-piston contacts.

Ford rated the street Boss at a modest 290hp and claimed that all of those ponies were on board by 5,800rpm. Most of the motoring cognoscenti at the time were highly skeptical of that rating and assumed Ford's modesty was more a soap to the insurance industry than an indication of a Boss' actual power production. And chances are good that a well-tuned street Boss actually did crank out significantly more horsepower; perhaps as much as 350hp (especially when the rev limiter was disconnected!). The work performed by the three groups of Ford stylists and engineers on the Boss project was not completed until well into the normal 1969 production sequence. As a result, Knudsen's best handling ever Mustang didn't hit the streets until April 17, 1969. By coincidence, the SCCA Trans-Am season that same year also didn't get underway until late Spring. And so it was that Boss 302s rolled into the showroom and onto starting grids at about the same time. The automotive world reeled from that one-two punch and, in many ways, was never the same again.

Preparing the 1969 Race Boss 302s

If there had ever been any doubt about Ford's corporate consternation at losing the 1968 Trans-Am title, it was fully erased when sponsorship plans were announced for the 1969 SCCA season. Ford fielded *two* fully financed teams on the circuit for 1969: one headed by Carroll Shelby and the other by Bud Moore who was being recalled to road racing from the NASCAR ranks for a second time. Ford was obviously not inclined to take any chances in 1969. With former Shelby team driver Jerry Titus campaigning Pontiac Trans-Ams for 1969, Shelby assigned driving responsibilities to road race regulars Peter Revson and Horst Kwech. Shelby's race cars came dressed in Ford corporate blue with the Boss Mustang's racey "C"

PJ drove number 15 Boss 302s like this one during the 1969 series. He scored the first Boss Trans-Am win at Michigan in May of 1969. Craft Collection

Father of the Boss: Larry Shinoda

Larry Shinoda is definitely a car guy. Born in that most automobile-oriented of West Coast hamlets, Los Angeles, Shinoda early on fell under the thrall of things powered by internal combustion engines. At the tender age of sixteen he was hanging out with hot rodders like Vic Edelbrock, Paul Schiefer, and Lou Barey. Soon he was making regular pilgrimages to both SoCal dry lake tracks and Bonneville with a series of Flathead- and Chryco-powered hot rods he'd built himself. When not actually wrenching on cars or racing them, young Shinoda was thinking about them. In time his formal education found him studying car design in earnest. He studied art and engineering at Pasadena City College and then went on to study technical illustration at the Douglas Aircraft Technical School before securing a degree from the Art Center College of Design in Los Angeles. His first stint at Ford came next in 1955 and coincided with a trip to the Indy 500 where he worked as a crew member on the winning open wheeled roadster (he later helped Roger Ward win the 500 in 1959 and 1962 as well). In 1956, Shinoda took his styling pens with him to Studebaker. When that car maker began its death spiral in the mid 1950s, Shinoda sent GM V.P. Harley Earl some Indy car sketches for review. That turned into a job at General Motors as a senior designer.

Shinoda's design work at GM is today well known by Corvette and Camaro partisans alike. It was his creative genius that produced both the original split window Sting Ray coupe and the even swoopier Mako Shark. Along the way, Shinoda had a hand in designing the Monza GT and the styling package that became the Z-28. He also played a role in designing Bruce McLaren's and Jim Hall's Can-Am dominating Group Seven race cars.

When Semon "Bunkie" Knudsen left GM for Ford, Larry Shinoda was one of the GM employees Knudsen made sure to take with him. Of Knudsen, Shinoda said at the time, "He's the greatest guy I know. While being very human, tolerant and sincere, he's most firm in action. He might be best described as an iron fist in a velvet glove." In addition to Shinoda's personal admiration for Knudsen, he also wholly supported his "build what we race; and race what we build" philosophy. Shinoda's first task at Ford was cooking up an all-new Mustang variant with a road race theme. As Ford fans worldwide are now aware, Shinoda's efforts produced the Boss 302. Shinoda was also instrumental in selecting the "Boss" name.

Perhaps less widely known is the fact that Shinoda was also responsible for designing the super swoopy Torino Talladegas and Cyclone Spoiler IIs that dominated the NASCAR high banks during 1969 and 1970 on their way to winning the factory backed "aero wars" (in which long-nosed Fords clinched 22 of the 35 superspeedway races contested). Shinoda's design group also penned the even sleeker King Cobra replacements for the Talladegas and IIs that would have come on line in 1970 had new Ford chief Lee Iacocca not savaged the corporate racing budget upon replacing Knudsen late in 1969.

Shinoda left Ford along with his corporate patron Knudsen late in 1969 and today owns his own automotive design firm based in Livonia, Michigan. Of late, Mr. Shinoda has begun to offer a cosmetic package for late model Corvettes, and an update version of the Boss package designed to work with the newest iterations of the Mustang line.

Larry Shinoda was one of the first General Motor employees that Bunkie Knudsen recruited after taking charge in Dearborn. Once ensconced in Ford's styling studio, Shinoda set to work on a series of high-performance cars that changed the face of racing. In that number were the Boss 302, the Boss 429, the Torino Talladega, the Cyclone Spoiler II and the stillborn King Cobra (pictured on the rendering to the left of Shinoda's hand in this 1969 Styling Studio shot). PPG Archives

The King Cobra Torino was slated to be the Talladega's 1970 season replacement. As penned by Larry Shinoda, the car featured a dramatically reconfigured beak that was designed to slice through the air. Boss 429 engines would have powered the cars to certain NASCAR victory had Lido Iacocca not slashed Ford's racing budget upon his ouster of Bunkie Knudsen. PPG Archives

Parnelli Jones' success at the season opener was marred by a mid-race shunt involving Horst Kwech's Shelby team Boss and a car parked in the infield that claimed a spectator's life. Craft Collection

stripes in sparkling white. Bud Moore's duo of drivers, Parnelli Jones and George Follmer, drove red, white, and black cars bearing the same white "C" stripes.

Unlike the first three notchback iterations of Ford Trans-Am cars, corporate Trans-Am race cars for 1969 were based on the far more aerodynamic "sportsroof" (fastback) unitbody. Street-going Boss cars bore an integral rear deck spoiler and chin spoiler specifically designed by Larry Shinoda to enhance the fastback's aerodynamics. Curiously, Shinoda's other aero add-on, the pedestal mounted trunk spoiler, was not used by Ford racers in 1969 (it did, however, venture into battle the following season).

Trans-Am pit roads were crowded with competitors in 1969. And so were the infield areas at every track on the circuit. Factory-backed racing was the reason. Craft Collection

Revson and Donohue fight for position. Craft Collection

In 1969, Fords were usually way out in front—for the first part of the season at least. Craft Collection

Beyond the aerodynamic cosmetics, the new-for-1969 race cars were the first Ford Trans-Am race cars assembled at Kar Kraft's Brighton, Michigan specialty fabrication facility rather than Shelby American. The first two prototypes didn't begin as Boss 302s, either! According to Ford archives, the first Boss 302 prototype (chassis number 9F02R112073) originally rolled off of the Dearborn assembly line as an "R" code, 428 Cobra Jet-equipped fastback. It was followed off of the River Rouge line by another "R" code chassis that carried a VIN just one digit higher than the first car's. Though assembled without sound deadener or seam sealer, the two cars were complete and running vehicles when their trip through the Rouge plant was over. This is quite unusual, as most of the Ford race cars were shipped from the factory unfinished. The white appearance of the unpainted bodies earned these cars the nickname, "bodies in white."

The two Cobra Jet-equipped fastbacks were shipped to the Kar Kraft plant, where they were completely dismantled.

Light is Right: Preparing the Chassis

Chassis preparation at Kar Kraft began with the elimination of every ounce of unnecessary weight. Then, as now, weight was the enemy of performance and great pains were taken to keep race weight as low as possible. Though the SCCA rules book for 1969 specified a race weight of 2,900lbs for any car competing in the five-liter category of the Trans-Am division, competition teams strived to trim as much weight as possible. The goal was to be well below the required SCCA minimum, and to return the weight advantageously. One of the goals was to redistribute the weight differential from the factory Boss 302's street safe front-biased (55.9% front/44.1%) weight to a race track friendly neutral or rear-biased weight balance. Also, by placing the returned weight low, lateral traction would increase and cornering speeds could rise. As a result, chassis engineers and race car fabricators made every attempt to carve pounds from the two "R" code chassis during their race preparation at Kar Kraft.

As mentioned, the bodies were already lightened on the assembly line where things like seam sealer and insulation were left off. Kar Kraft began reducing weight by removing flanges and structural supports not needed for race duty. Components required by the official rules book but unnecessary for fast lap times were then treated to a drill bit induced weight loss program that left them resembling a piece of metallic Swiss cheese. So much structural rigidity was removed from the door window regulator mechanisms (complete with shortened cranks that saved a few more ounces) that "R Model" style lift straps were needed to hoist the cars' side glass. Even the factory glass was replaced with tempered glass replacements that were significantly thinner (and lighter) than stock.

More weight was saved by using special "drag race" style fenders, door, and hoods stamped from thinner than stock sheet metal. It's also been rumored that at least some parts of the car's chassis were trimmed with a dip in the acid vat. Although acid dipping was specifically prohibited by the SCCA rules book, Roger Penske perfected the art of chemical milling in 1968 when his team Camaros were put on a crash diet in order to catch their Shelby Team rivals.

As the process' name suggests, chemical milling involved immersing a part (or entire unitbody) in a tank of caustics for a length of time calculated to remove a specific amount of mass. When performed correctly, an amazing amount of weight could be trimmed. *Sports Car Graphic* magazine (in a September 1969 article about the illegal but open practice) estimated that between 200 and 300lb of body and engine mass could be leached from a Trans-Am car's total weight if every possible component (including the car's engine block) was treated to a trip through the acid tank. The article in question even featured

Although slow by NASCAR standards, Trans-Am series pit stops were still expeditious affairs. Here George Follmer's Boss receives service at Lime Rock. Craft Collection

snap shots of a unitbody being lowered into the acid vat that just coincidentally looked a lot like one of Penske's Camaros. Timing was critical during chemical milling as parts left too long in the solution became dangerously or even worse, visibly weakened. It's been widely reported, for example, that during the 1970 Trans-Am season, the unitbody of Sam Posey's Dodge backed Challenger T/A was dipped to the point that its roof panel waved in the wind like a flag. According to the story, the team became so concerned about a visit from the SCCA technical crew at one stop on the circuit, that a rental Challenger coupe was secured from a local agency and then returned as a convertible—its stock roof having been "borrowed" for use on Posey's race car.

Once sufficiently lightened (by whatever means), the two prototype Boss race cars were then reinforced with a jungle gym's worth of roll cage tubes. As built by Kar Kraft, a Boss' cage consisted of a six-point, diagonally braced cage that was tied directly to suspension mounting points for superior rigidity. Those suspension points themselves were also the objects of fabricator attention. They received both new geometry and additional bracing with an eye towards stiffening the chassis' resolve under cornering loads.

Lastly, Kar Kraft fabricators made a few modifications to improve aerodynamics by cleaning up the sportsroof profile before the suspension components were bolted into place. Using aerodynamics to improve performance was cutting edge for the time but, as with acid dipping, not all of these steps were, er, strictly legal. Take, for example, the one-inch slice of metal that was sectioned out of the two Kar Kraft chassis' radiator support. When mated with inner en-

Lime Rock was the second stop on the 1969 Trans-Am trail. Because that race fell on Labor Day, Peter Revson, away at a little known race in Indianapolis, was replaced by Sam Posey. As it turned out, Lime Rock was Posey's home track. He put his familiarity with that sinuous Connecticut circuit to good use and made it two for two for Boss 302s. Posey scored Shelby American's last team victory for the Ford Motor Company. Craft Collection

gine box aprons that were commensurately tapered down from the firewall, the whole of the Boss front sheet metal profile could be lowered significantly. And that, not coincidentally, greatly enhanced the finished cars' aerodynamic silhouette. So, too, did the sectioned (and therefore raised) subframe boxes that were also fitted to the chassis. As with acid dipping, these subtle changes went undetected by SCCA tech inspectors.

Brakes, Suspension, and Handling

Once chassis modifications were completed, heavy duty racing underpinnings were fitted. At the bow, beefy spindle forgings served as the suspension's central focus. They were articulated to the chassis via reinforced upper and lower control arms that had their geometry altered to improve camber change under lateral acceleration. Heavy duty hubs, huge 11.96in diameter ventilated rotors, and fixed four-piston Kelsey-Hayes calipers lifted from the Lincoln line came next. Their extra weight was translated to the chassis via a set of track-specific (though always kidney pummeling) coil springs. Two-way adjustable Koni race shocks, a wrist-thick sway bar fitted with adjustable heim joint ends, fully metallic suspension bushings, and a heavy-duty drag link/tie rod assembly rounded out the front suspension.

Before moving aft, the Kar Kraft crew also outfitted the front half of the chassis with a race-proven export brace and Monte Carlo bar combination for a bit of flex resistance in the twisties. A NASCAR-style (but unique to Kar Kraft) "full floater" 9in differential was mounted under the rear frame rails. It was connected to the rest of the chassis via a set of override traction bars (quite similar to those first devised by Shelby American for 1965 GT350 duty), a pair of race-rate leaves, and a heim joint-equipped Watts link designed to hamper lateral movement. A smaller version of the front sway bar was also part of the package as were a duo of Koni race shocks. Another set of Kelsey-Hayes four-piston calipers and an equally massive pair of ventilated discs were employed to complement the speed scrubbers mounted forward. The whole chassis rolled on magnesium rims that measured 15x8, and carried sticky 5.00x11.30 (front) and 6.00x12.30 (rear) treaded race tires. Special 5in lug nuts kept those rims and tires working in close harmony with the rest of the suspension. Dressed for war, the first race Boss hunkered over the pavement at a much more intimate distance than a street car. The 1in front and 3.5in rear ride height the Kar Kraft Trans-Am chassis measured out at, while adequate for glass smooth race course duty, would have collected just about every manhole cover encountered on a regular surface street.

A number of replica Trans-Am Bosses have been built for service in the vintage ranks. Although not bona fide Kar Kraft cars, cars like this Bud Moore team replica still help recreate the thunder and glory of the racing Boss 302s.

The net results of Kar Kraft's chassis modifications were internal organ-displacing cornering speeds and tarmac-shredding stopping ability. A Bud Moore Boss 302 road tested by *Road & Track* magazine in 1971 was capable of pulling 1.1 times the force of gravity on a 100ft skid pad. When the anchor was thrown out via the non-power assisted brake pedal, a full stop from 80mph was possible in just 180ft. Compare those statistics with the .73G skid pad performance and 292ft stopping distance of a street Boss and you'll understand why the *Road & Track* guys wrote that they didn't know how anyone could build a better Trans-Am car.

Upon completion, the two "R" code prototype Trans-Am Bosses were dispatched to the Shelby and Moore teams for evaluation. Each ultimately saw duty during the 1969 season. In addition to those two chassis, Kar Kraft was also responsible for the complete construction of one other 1969 Boss 302 competition car, the spectacularly beautiful black and gold chassis that was slated for use in the NASCAR Grand American series by Knudsen's favorite race car mechanic, Smokey Yunick.

Although just about every Boss 302 that raced for the factory teams during the 1969–71 seasons is commonly referred to as a Kar Kraft car, Kar Kraft was only completely responsible for the first two prototypes and Yunick's car. After the two "R" code cars were completed, six more fastback Mustangs were diverted from the Dearborn assembly line to the Kar Kraft facility. Unlike the first two 428-powered cars that Kar Kraft had converted to Boss race cars, the next series of Mustangs to arrive in Brighton were all driven off the line powered by 351W engines that were backed up with four-speed transmissions. As a result, the next crop of "Boss 302" Trans-Am race cars all carried "M" code VIN numbers. At Kar Kraft, the six chassis (three slated for use by the Shelby team, and three more destined to be delivered to Bud Moore in South Carolina), received all of the suspension modifications that had been fitted to the original prototype Boss chassis but little more. Which is to say that most other race car conversion work, including roll cage construction, was completed by the individual teams. As a result, there were a great many variations and individual differences between those six cars (and the others that followed, excepting the Yunick chassis) even though all have collectively come to be called "Kar Kraft" cars.

All told, ten "M" code sportsroofs were pulled off of the Rouge line to serve as the starting point for "regular production" Trans-Am cars, according to Ed Ludtke's Trans-Am Car Registry. Three "bodies in white" (bare metal, unserialized unitbodies pulled from the assembly line just after being "bucked"), were built to Trans-Am specs by Kar Kraft and Bud Moore for use during the 1971 season and one bare "Kar Kraft" chassis went unused until being restored to Trans-Am specs in the late 1990s.

When the six original Trans-Am chassis had been converted to race configuration, they were sorted out by their teams in preparation for the upcoming 1969 season. Power for those time trials was provided by a race engine package selected as a result of R&D work conducted late in 1968. At that time, three engines were under consideration for race duty during the upcoming Trans-Am season: the Tunnel Port 302, a Gurney Westlake-headed version of the 302, and the Cleveland-headed version of the 302 that became the Boss. Shelby American driver Horst Kwech tested each of the three engines in one of the team's 1968 notchback Trans-Am cars on the late great Riverside road course in California. At that time, the Gurney engine proved to be the fastest, the Tunnel Port the slowest, with the Boss-equipped car somewhere in between. Ultimately, the Tunnel Port and Gurney engines were eliminated from consideration based both on performance and Ford's ability to produce their components in numbers necessary to placate the sanctioning body.

During the Boss' racing development many different induction packages were tested. Some of the tests conducted by Shelby American, employed Boss motors outfitted with a quartet of Weber carburetors. Ford also contemplated casting up a dual barrel carburetor setup that featured offset Holleys mounted Z-28-fashion on a high rise alloy intake. That intake was referred to by the cast logo "Cross Boss" (later used to identify a special Weber-like In-line carburetor and intake developed by Ford for the 1970 season). When testing proved it to be fraught with fuel reversion problems, Ford engineers settled upon a more conventional 8V induction system that mounted twin Holleys in-line on a dual-plane manifold. Ultimately, a twin 1050 Holley Dominator intake system was used for the majority of the 1969 season. Just short of its 8,500rpm redline, a race spec Boss motor could be relied upon for upwards of 450hp. Both Shelby and Bud Moore got the pieces necessary to build their race motors directly from Ford's Engine and Foundry Division and were responsible for their team engines' final assembly and development. When fitted to a properly setup Kar Kraft developed chassis new lap records were just a stab of the throttle away. Recapture of the Trans-Am title seemed a foregone conclusion as the 1969 Motorsports year dawned.

The 1969 Trans-Am Season
Unlike the preceding Trans-Am seasons, neither the Daytona Continental nor the Sebring endurance races were part of the 1969 Trans-Am schedule. Instead, the season opener was slated for May 18th at Michigan International Speedway. But that didn't mean that Boss teams waited until the Spring to seek their first blood. Instead of indolence, Ford opted to blood the Boss 302 at the first available opportunity. And that turned out to be the Florida Citrus 250 at

Factory Trans-Am Boss 302s

From Ed Ludtke's *Trans-Am Car Registry*

This listing is taken directly from the *Boss 302 Registry* lists all of the 1969 and 1970 Mustang Boss 302 Trans Am race cars prepared by Kar Kraft as well as the car's last known condition.

Shelby Team Prototype/Test Car, #9F02R1120731969

Became main team car and was wrecked at St. Jovite 8/3/69. The car was repaired and saw limited action the remainder of the 1969 season. Was raced as an independent in 1971, 1972, then went to Mexico. Later returned to the United States and won A-Sedan Championship in 1979. Status (1995): unrestored, still in A-Sedan configuration.

Bud Moore Team Prototype/Test Car, #9F02R1120741969

Became main team car and was the first Boss 302 used in competition. Driven by Parnelli Jones at the Daytona Citrus 250 race 2/21/69. Became George Follmer's 1969 team car and was wrecked at St. Jovite 8/3/69. Car remained in wrecked condition until 1990. Status (1995): restored.

Bud Moore Team Car, #9F02M1486231969

Became George Follmer's main car after Follmer's first car was wrecked at St. Jovite. This car also became wrecked from a broken wheel at Riverside, California, while leading that final race of the season. Car remained in wrecked condition until 1992. Status (1995): undergoing restoration.

Bud Moore Team Car, #9F02M1486241969

Never raced in the United States, was exported while still brand new to Australia and was raced extensively by Allan Moffat. Returned to United States in 1990. Status (1995): As raced in Australia, unrestored.

Bud Moore Team Car, #9F02M1486251969

Specific history and 1995 status unknown.

Team Car Built For Smokey Yunick, #9F02M1486261969

Never raced in the Trans-Am series. Was converted by Smokey to Grand National specs and raced in NASCAR's Baby Grand series by Bunkie Blackburn. Later raced by Ed Rose as a short track stock car. Status (1995): restored

1969 Shelby Team Car, #9F02M148627

Driven by Horst Kwech at first Trans-Am race at MIS and wrecked there. Car was stored at Holman & Moody and eventually parted out. Status (1995): unknown.

Shelby Team Car, #9F02M1486281969

Driven by Sam Posey and Dan Gurney. Became Bud Moore backup car in 1970. Sold to Tony DeLorenzo of Troy Promotions and raced by him in 1972. Stored until 1986. Status (1995): restored and vintage raced.

Shelby Team Car, #9F02M1486291969

Driven by Peter Revson and Horst Kwech, wrecked at St. Jovite 8/3/69. Status (1995): unknown.

Bud Moore Team Car, #9F02M212777

Driven by Parnelli Jones. Became 1970 backup car. Sold to Tony DeLorenzo of Troy Promotions and raced by him in 1972. Later raced by George Gunlock. Stored until 1982. Status (1995): authentic original condition, unrestored, and vintage raced.

Bud Moore Main Team Car, #9F02M2127751970

Originally built with unique three-link rear suspension and was Parnelli Jones' primary car. Won 1970 Trans-Am championship on 9/20/70. Raced by Parnelli Jones and George Follmer for Bud Moore team in first half of 1971. Sold to someone in Mexico later that year. Status (1995): unknown.

Bud Moore Main Team Car, #9F02M2127761970

Originally built with unique three-link rear suspension and was George Follmer's primary car. Sold to Warren Tope in 1970. Raced by Tope until 1973 and won the 1971 A-Sedan championship. Raced by Bill Clawson from 1973 to 1975. Stored until 1980. Status (1995): restored and vintage raced.

Bud Moore Team Car (Body In White), #11971

Body and chassis left over from 1970 Ford Trans-Am program and given to Bud Moore. Completed for 1971 Trans-Am season and driven by Peter Gregg and George Follmer. Raced by Marshall Robbins in 1972 and Phil Bartlett in 1976-1981. Sold to Ed Ludtke in 1982 and Fossil Motorsports in 1988. Status (1995): restored and vintage raced.

Bud Moore Team Car (Body In White), #21971

Body and chassis left over from 1970 Ford Trans-Am program and given to Bud Moore. Completed for 1971 Trans-Am season and driven by Peter Gregg and George Follmer. Raced by Marshall Robbins in 1972-1973. Raced by Mike Lovejoy in 1974-1976. Stored until 1989. Status (1995): unrestored original condition.

Bud Moore Team Car (Body In White), #31971

Body and chassis left over from 1970 Ford Trans-Am program and given to Bud Moore. Completed for 1971 Trans-Am season and driven by Peter Gregg and George Follmer. Sold to Dan Doughtrey and Wendell Scott in 1974 and raced by them until 1977. Stored until 1988. Status (1995): unrestored.

Body In White #4

Body and chassis left over from 1970 Ford Trans-Am program and given to Bud Moore. Never completed and no history. Sold by Bud Moore as bare shell to Paul Pettey. Sold to Curtis Jackson in 1991. Status (1995): unbuilt shell.

And that turned out to be the Florida Citrus 250 at Daytona in February. The Citrus 250 was one of the first stops on the NASCAR Grand American tour in 1969. Similar in concept to the SCCA's Trans-Am series, the Grand American tour (often called the Baby Grand tour) featured American pony cars and circle track competition.

Bud Moore and Parnelli Jones came to Daytona with a red, black, and white Boss 302 that had been specially prepped for the 250. Though lacking a set of Shinoda inspired "C" stripes and wearing a set of non-SCCA spec headlight covers (for improved aerodynamics down the Big D's back stretch) the car was essentially the same package that was scheduled to make its Trans-Am debut at Michigan. During practice, Jones was one of the fastest cars on the track and though the number 15 Boss missed the pole (to Don Yenko's ex-Smokey Yunick 1968 Camaro) by a fraction of a second on pole qualifying day, PJ was clocked as much as a second faster than the rest of the field during general qualifying. Jones put that speed advantage to good use when the green flag fell and simply drove away from Yenko and the rest of the field. Unfortunately, transmission failure on lap 20 resulted in a DNF for the Boss' first outing. Even so, Jones' performance promised happier days ahead.

At Michigan International Speedway three months later, Moore team cars dominated qualifying. When the green flag fell Parnelli Jones and George Follmer owned both halves of the pole position. Shelby drivers Horst Kwech and Peter Revson weren't far behind on the second and third rows, respectively. Though snow, rain, and hail fell on the Irish Hills of Michigan throughout the race, the on track action was close. At race's end, it was Jones who led the pack across the stripe for the Boss 302's first win (although a scoring error had initially placed him fourth). That triumph was marred by a fatal shunt involving Kwech's Shelby team Boss and an infield spectator. Mechanical DNFs scored by Follmer and Revson were also worrisome.

The road course in Lime Rock, Connecticut was

Boss 302s were only available in a limited number of hues in 1969. Bright Yellow was the brightest of the bunch.

next up on the 1969 schedule. With Jones, Revson, and Follmer at another race in Indiana on that particular Memorial Day weekend, Ford drafted Swede Savage, Sam Posey, and John Cannon to take their respective places. Kwech's blue Boss was the car to beat in the early running and led until brake failure allowed Posey to slip past. Posey led for the balance of the contest challenged only by Savage for first place honors. Two races into the season, it looked like Ford would have a very good year in Trans-Am racing.

Jones and his Ford teammates were back from Indy by the June 8th Trans-Am stop at Mid-Ohio. As was his habit, Jones charged to the front of the pack and rubbed fenders with Mark Donohue (in the Penske Z-28) for most of the race (to the extent that he'd collected a rough driving complaint not far into the race). Jones led until his first fuel stop at which time Ronnie Bucknam, Penske's other driver, assumed the lead. At race's end, it was Bucknam, Jones, Follmer, and Revson who crossed the stripe in that order. Though not a Blue Oval win, the finish added to Ford's growing lead in the all important points race.

Bridgehampton brought the Ford team another Boss blowout in June. Though Donohue's Camaro qualified first, engine incontinence on the warm-up lap put him out of contention in a last place starting grid back up car berth at the green flag. In his absence, Jones and Follmer swapped the lead for the first 30 laps. Follmer and his number 16 "C" striped Mustang took the checkered flag first to make it three wins in four races for Ford.

The gods of racing can be fickle indeed. Though Ford seemed on the verge of humiliating the competition on the SCCA circuit at just about every stop on the 12-race Trans-Am tour, mechanical woes kept Ford drivers from both of the top two positions for the first time at Loudon, New Hampshire. Even so, Revson's third and Follmer's fourth still stretched out Ford's overall points lead as the circuit crossed the border on the way to a stop at St. Jovite, Canada. In retrospect, it was one trip that the Shelby and Ford teams should never have taken.

The seventh race of the season began with the expected charge to the front by Jones and Follmer. With Donohue in the hunt for the lead, the first eight laps were a flurry of close quarters action, with the three cars never holding their positions for more than a lap. Unfortunately for the Ford teams, on lap nine Jones' number 16 Boss coasted off course with a transmission rendered immobile by a jammed shifter linkage. Five circuits later, George Follmer ventilated the oil pan in his Boss and spun in the resulting goo. The car came to rest just around a blind curve and caused a pile-up that took out Kwech's Shelby Boss and nearly a dozen other cars. One of the last to the wreck was Revson in the second Shelby car. He hit the pile of shredded cars with so much force that his blue number one car jumped over one car and came to rest atop another. Not realiz-

Boss engines were visual tour de forces and came factory-dressed in chrome valve covers and chrome "spoked" air cleaner lids. Their performance was sparkling, too. Note the vacuum diaphragm "flapper valve" mounted on the side of the air cleaner that was the only type of Ram Air induction available on Boss 302s in 1969.

ing that he was on top of another car when the dust settled, Revson crawled from his car and fell to the tarmac, injuring his shoulder.

Of far more concern to the Ford effort than Revson's bruised shoulder was the intensive care status of the three team Bosses. As it turned out, they were all nearly total losses. The Shelby and Ford teams spent the two short weeks before the next race trying to cobble up the parts to field complete teams. The fact that Donohue's win at St. Jovite had placed Ford behind in the championship race for the first time added extra urgency to those efforts as the series crossed the country to run at Laguna Seca in Monterey.

All seemed well with the Ford effort during qualifying as Jones and Follmer once again were the fastest while running for the pole. They reinforced that fact by leading the early stages of the two-and-a-half-hour event. But on lap 44, the Jones car's differential expired in a cloud of foul smelling smoke. Follmer took up the lead for the next few laps until he, too, was sidelined with mechanical gremlins. Donohue inherited the lead and went on to win, further extending the Bow Tie Division's points lead.

The odds clearly seemed stacked against the Ford teams as they pulled into Pacific Raceways Park at Kent, Washington in September. Simple math indicated that team Bosses would have to finish first and second at all of the remaining three races if Ford was to regain the Trans-Am title in 1969. With just one race-worthy chassis remaining between the two corporate Ford efforts, that seemed an unlikely prospect. Jones put the remaining car to good use by leading the first 74 laps of the 300-mile affair. The impossible seemed to be within reach when Donohue's Sunoco Camaro

45

Boss 302s in NASCAR Racing

While the exploits of Parnelli Jones, George Follmer, and the other SCCA Boss 302 drivers are widely known and warmly remembered, the triumphs scored by a contemporary group of Boss Mustang racers is mostly forgotten these days. We refer, of course, to the triumphs scored by Boss 302 Mustangs in the NASCAR ranks. But, fact of the matter is, Boss 302 drivers made their mark in roundy round, good ol' boy circles, too.

The venue was NASCAR's "pony car" based Grand American (GA) series that ran as sort of an under card to the big league Grand National (GN) stock car circuit. Cars on the GA circuit usually ran on the Saturday preceding a GN event on the same oval track that the big boys would be rocketing around the very next day (much the same as cars on the Busch circuit do today). The field of a GA race was usually composed of a mixture of factory-backed and independent teams, with the sponsored cars often being driven by one of the Grand National ranks heroes.

As most Boss 302 fans will recall, the very first competitive laps taken by a Trans-Am spec Boss car occurred in February 1969 in the Citrus 250 when Bud Moore and Parnelli Jones debuted a red, white, and black Boss 302. That particular car featured a NASCAR-style, fender-mounted fuel fill arrangement and ran without a set of "C" stripes in its one-time appearance on the Grand American circuit.

When Parnelli Jones and Bud Moore left Daytona on their way to full-time work in the SCCA ranks, others took over the reins of the Boss 302 Grand American competition. Smokey Yunick was in that number. Like the Shelby and Moore teams, Yunick had received a prototype Boss 302 in running condition straight from the race fabrication Kar Kraft concern in Michigan. Smokey's car was arguably the prettiest of the trio and came decked out in his trademark black and gold racing livery. As delivered, the car carried a set of gold "C" stripes and was built to full Trans-Am specs.

By 1969 Yunick had tapered off his racing activities considerably and spent the majority of his time working on engine performance at his Daytona Beach shop. That work included perfecting the Boss 429 induction system intended for high bank use on the Grand National circuit. The dyno that Ford setup for Yunick in his Best Damn Garage in Town also helped Yunick conduct his R&D work on a wide variety of other Ford engines.

In addition to the Boss 302 sent south to Daytona Beach, Smokey was also assigned a black and gold Boss 429-powered Torino Talladega. Chances are, Smokey might have limited himself to a handful of racing forays with the Talladega had it not been for a driver's strike that sidelined most of the NASCAR regulars just before the inaugural running of the Alabama 500 in Talladega. The impetus for that strike was a tire wear problem that drivers claimed made the new superspeedway too dangerous to run on. Ford driver Richard Petty headed up the driver's union that led the boycott and just about every Ford and Chryco driver followed his Talladega out of the speedway prior to race day. That posed a problem for Ford boss Knudsen, whose father was slated to be inducted into the International Motorsports Hall of Fame as part of the race week's festivities. As head of the Ford Motor Company, Knudsen was desperate to have a Ford race car running at the track that weekend. And that's where Smokey Yunick came in. Though Smokey, like the rest of the NASCAR regulars refused to run his Talladega in the Grand National race, he did agree, on a moment's notice, to enter his black and gold Boss 302 in the Grand American race scheduled to run before the main race on Sunday. As Smokey tells it today, he tapped journeyman NASCAR driver Bunkie Blackburn to drive the car and put together a crew to support the car at Talladega just hours before the race.

Race preparation of the Boss involved a number of "Smokey specific" chassis modifications that were quite different than the suspension changes designed by the engineers at Kar Kraft. Take, for example, Smokey's use of an outboard (of the frame rail) Australian Falcon steering box. When connected by a specially lengthened drag link to an idler arm that was also moved outboard of its normal mounting point, the new steering arrangement created lots of extra space for a unique header system Yunick had cooked up. Even more space for that serpentine assemblage of stainless steel tubing was created by the reverse-wound starter that Yunick also bolted on the Boss (to a special bell housing that placed the starter through the firewall).

Other suspension changes conducted on the Boss included the unique combination of negative and positive camber that was dialed into the front tires with an eye towards making left turns on Talladega's 33 degree banking. Double centered,

Smokey Yunick's second stint with Ford came during the 1969 NASCAR and SCCA seasons. Smokey helped perfect the Boss 429 engine for Grand National use and he also was assigned a fleet of Torino Talladegas and Boss 302s for selected racing use. Smokey's first outing with a Boss 302 came at Talladega in September of 1969. His black and gold Kar Kraft Boss car sat on the pole at the Grand American race that ran in conjunction with the inaugural Grand National event at Bill France's new superspeedway. Here Yunick (right) confers with Boss driver Bunkie Blackburn (with glasses) and a team member just before the race was run. **Daytona Racing Archives**

stamped steel wheels (with special cooling scallops) were also used on the Boss instead of the magnesium rims common to most other racing Boss 302s.

On race day, Smokey's car showed up wearing his traditional number 13 identification and sans its Kar Kraft applied gold "C" stripes. When the green flag fell, Blackburn quickly translated his pole starting position into a commanding lead. According to Yunick, at one point the black and gold Boss was fully two laps ahead of the rest of the Grand American field. But defeat was snatched from the jaws of victory by a pushrod that worked its way completely through a rocker arm during the race. As a result of that failure, Yunick's Boss 302 was a DNF that day, and ultimately never ran again in either Grand American or SCCA trim.

Besides Yunick's Kar Kraft built Boss, a number of Holman & Moody (H&M) prepared Boss 302s also took part in the Grand American series in 1969 and 1970. Without a doubt, the most successful of those cars was the red number 49 Boss-powered fastback that both David Pearson and Bobby Allison campaigned in 1970–71. Based on the few existing pictures of that car, it appears that it shared many of the Grand National developed construction features that Holman & Moody had perfected for full-sized stock car competition. As a result, it probably began life as a body in white chassis specially modified in H&M's Charlotte racing factory.

The Holman & Moody-built Boss enjoyed its greatest success in 1971 when Bobby Allison drove it to victory in a Grand National race. That's right, a Grand National—not a Grand American—race. Allison had taken Pearson's place as H&M team driver upon the "Silver Fox's" departure early in 1971 and the gold and red, number 12 Cyclones he drove for Ralph Moody and John Holman were more than a little successful. When not behind the wheel of a Boss 429-powered Mercury, Allison drove the H&M-prepped Boss 302 at selected events. Late in 1971, when NASCAR officials decided to combine the Grand American and Grand National divisions at a handful of short track races, Allison opted to run his Mustang instead of the larger and more cumbersome Mercury. That decision paid off at the Myers Brothers Memorial race held at Bowman Gray Stadium in August. Allison qualified his Boss 302 second that day against pole sitter Richard Petty's full sized 426 Hemi-powered Road Runner. When the green flag was shown, Allison tagged along behind Petty until lap 113 of the event when he assumed the lead. Once out in front, he never looked back and on lap 250 he scored the first (and only) Grand National stock car victory for a Boss 302 Mustang.

Though Allison backed up that run with a solid second place Boss 302 finish at the next event (the West Virginia 500), he and his Boss never visited victory lane again and soon NASCAR discontinued the experiment of combining the two series. Interestingly, though Allison won the Myers Brothers event without protest, the sanctioning body does not recognize that win as one of Allison's Grand National victories. And so, today Allison is recorded as having scored 84 NASCAR wins instead of the 85 he actually won.

Delivered to Yunick in Trans-Am trim, Smokey's Boss underwent a number of changes before it suited his tastes. In that number were the removal of the car's gold "C" stripes, the repainting of all top surfaces in gold, and the application of Smokey's trademark number 13 identification. Note the specially scalloped NASCAR styled rims that were used on the car. During the race, Smokey's Boss was so fast that it at one point had lapped the field twice. Unfortunately, valve train trouble sidelined the car before the checkered flag. Daytona Racing Archives

valve and a last lap flat tire (that Jones ignored by driving on the rim) ultimately conspired to allow Penske's other Z-28 to cross the line first. Jones' sparks flying second place finish still left Ford at a 10-point deficit with just two races left to be run.

Even though at a seemingly insurmountable disadvantage, Ford teams didn't just throw in the towel. At the very next venue, for example, Jones, (whose motto was, "If you're under control; you're not going fast enough") blitzed the field with a qualifying run at Sears Point that was two full seconds faster than the field. When the starter released the grid, Jones charged to the lead and bested the field for the first 69 of the 80 laps that made up the event. No one could touch Jones on the track that day, but that's not where the race was won. Though Jones only made two stops for service during the event, he spent a total of 60sec parked on pit road. Donohue's THREE stops, on the other hand, consumed just 34.9sec. As hard as he tried once back on the track, Jones was never able to overcome that deficit and he lost to the Penske car by 2.17sec. When the race was over, Chevrolet had won its second straight Trans-Am title.

The last race of the 1969 season was held at Riverside International Raceway in October. With the 1969 season no longer in doubt, the only thing left to race for was bragging rights. And Ford went after those with an expanded five-car effort that included Indy ace Al Unser in a third Bud Moore car. Jones set a new Riverside lap record to sit on the pole for the 250-mile event and as usual charged out from the flag to take the early lead. Unfortunately, the pressures of the season had created a personal enmity between Jones and Donohue. At Riverside those tensions came to a head in the form of some NASCAR-style fender rubbing. The "love taps" that Jones and Donohue delivered to each other would wind up hurting Jones' Ford more than Donohue's Chevy. When George Follmer also succumbed to mechanical problems, Donohue's battered but still running Z-28 crossed the line first. Though Ford's new Boss 302 had been fast all season, minor mechanical woes and the season sabotaging shunt at St. Jovite had conspired to deny them the Trans-Am championship. Things would be different in 1970.

The Boss 302 Hits the Boulevards

While Parnelli Jones, George Follmer, Horst Kwech, and Peter Revson were doing battle on road courses all across America, Boss 302s were competing at an even greater number of North American venues: the showroom sales floor. And sales were, after all, the impetus for factory backed motorsports in the first place.

Knudsen and his fellow Ford executives were banking on the excitement generated by the Boss teams' on-track efforts. The phrase "race on Sunday; sell on Monday" never had more cachet than in the early months of 1969. Records show that 299,824

The rear deck spoilers cooked up by Larry Shinoda for the Boss (and regular Mustang line) in 1969 were molded from lightweight fiberglass that tended to bow in the middle over time. Although not originally part of Shinoda's styling design, sagging spoilers are today a signature feature of just about every correctly restored 1969 Boss 302.

Mustangs of all stripes were built during the 1969 model year. In that number were the 1,628 street going (G engine code) Boss 302s that were the direct result of the SCCA's homologation rules for the 1969 season. Their appearance on the sales floor in concert with the Boss teams' success on the Trans-Am circuit created a sensation. When the cars first began to show up on dealer's lots in April of 1969 (as 1969 ½ models), sales floor traffic went through the roof.

A new Boss 302 was essentially a package option. The standard engine was a high-revving version of the cant-valved Boss 302 equipped with a cobby solid lifter cam, a true dual exhaust system, and a huge 780cfm Holley carburetor bolted to an aluminum high-rise intake manifold. Ford hadn't forgotten the impact of underhood chrome while cooking up the Street Boss package. Boss buyers of 1969 had only to pop the hood to wow the crowd at the local cruise spot with chrome-plated valve covers and air cleaner bonnet. Other Boss underhood appointments included a factory installed rev limiter (that was only semi-successful in nipping engine-related warranty work in the bud), a vacuum operated pop-off valve that allowed extra air to the engine under full throttle, and a set of racy "Boss 302" valve cover decals.

In a day when Detroit car makers usually paid little attention to underhood appearances, a new Boss 302's engine bay was a visual tour de force—and that was before the free revving little engine was summoned into life. Ticking over at idle speed, the engine

Although sybaritic comfort was just an option box away for Boss buyers in 1969, most Trans-Am Mustangs built that year came with base black interiors. The chrome-plated Ford shifters in those cars provided just about the only available interior flash. When those shifters were rowed through the gears with vigor, most Boss owners didn't seem to mind the lack of interior appointments.

Acapulco Blue was one of the four colors available for 1969 Boss 302s. It looked particularly attractive when complemented by a Boss car's tape and black out treatment.

made all the right noises; the solid lifters ticked in mechanical harmony, the four-throated Holley made a delightful "whaaw" when jabbed by an enthusiastic right foot, and the dual exhaust rumbled with basso profundo authority.

The basic mechanical elements of the street Boss package were surprisingly similar to the hard parts found under Parnelli Jones' roundel-covered hood. A beefy four-bolt main block served as the engine's starting point. And that "C8 or C9" part numbered casting came fitted with a sturdy forged steel, cross-drilled crank, just like the race cars. Forged pistons worked in concert with that crank and an octet of forged rod castings equipped with beefy 3/8in rod bolts. When the cam was on its base circles and the 2.23in/1.71in valves were fully closed, the domed pistons in each cylinder squeezed combustibles into quench-shaped combustion chambers at the rate of 10.5:1. Rev cups, damper-equipped springs, and factory installed guide plates all did their part to stabilize the stamped steel, 1.73:1 rocker arms and the rest of the valvetrain.

Other bits and pieces of the street Boss engine package included a vacuum advance, dual point distributor, an underdrive alternator pulley, a baffled five-quart oil pan with a sporty, chrome-plated dip stick, a main cap mounted windage tray, a Carter high-volume fuel pump, a high capacity oil pump, a five blade "flex" type fan and a balanced water pump impeller. A horsepower robbing air injection style smog system was also part of the underhood landscape. At least until a new Boss buyer had time to break out his socket set.

Ford had become quite conservative with its horsepower rating by 1969 for fear that reporting true figures would produce sales-killing insurance penalties. As a result, the new engine was rated conservatively at 290hp. Torque was calculated at 290lb-ft.

Two versions of Ford's venerable Top Loader four speed but no automatic transmission were available for the Boss 302. A wide-ratio four-speed (2.78:1 first gear) came as standard Boss equipment but those in search of shorter shifting could order up a close-ratio box (2.32:1 first gear). In either event, a Ford shifter and linkage was installed to row the box through the gears. Unless ordered otherwise, a 1969 Boss came outfitted with a nodular, case-mounted, 3.50:1 ring

All Boss 302s were fitted with argent-centered 15x7 Magnum 500 rims for 1969. Chrome lug nuts and short (as compared to 1971 and later Magnum rims) "running horse" center caps were also part of that flashy package.

1969 Boss 302 Road Test Results

Car and Driver, June 1969
- zero to 40 ....3.2sec
- 60 ....6.0sec
- 80 ....10.0sec
- 100 ....15.2sec
- ¼ mile 14.57sec@97.57mph

Sports Car Graphic, June 1969:
- zero to 40 ....4.3sec
- 60 ....6.0sec
- 80 ....10.9sec
- 100. ....16.8sec
- ¼ mile 15.00sec@96.00mp

1969 Boss cars could be ordered up with plush deluxe interiors like the one in this Wimbledon White car. Comfort weave covered high-back buckets, padded door panels, and a simulated teak dash and door panels were part of that interior option. This car also sports an in-dash tachometer, a floor-length console, and a passenger side "rallye clock."

and pinion and nearly indestructible 31-spline axles. A 3.91 and 4.30:1 gear set was available for racier performance and was accompanied with an auxiliary oil cooler. Traction-Lok was also an option.

According to Boss sales literature a "competition-type" suspension was installed under all street Bosses. What that translated into was high rate front (350lb/in) and rear (150lb/in) springs, an increased (.85in) front sway bar, staggered rear shocks, and increased rate Gabriel shock absorbers. Power steering was an option and a standard 16:1 steering box was used to endure quick steering response. Floating single-piston calipers acted on 11.3in discs at the bow and 10in shoes and drums were mounted rear. Power assist was a standard part of the Boss braking system. The whole Boss package rolled on flashy, argent-center Magnum 500 rims and racy F60x15 super wide oval Fiberglas belted tires.

While Ford offered only a few performance options for the Boss, there were dozens of choices for the interior. As the Boss 302 sales brochure trumpeted, "The Boss offers a choice of interiors that's unequaled in its field." In point of fact, every 1969 Mustang interior option was available. Those choices began with hue, and no fewer than six different shades of vinyl could be ordered in a Boss 302's cockpit. The next big choice was between high- and low-backed bucket seats. Low-backed buckets came standard, but racier high-backs were just an order box away. And then there was the choice between plain vinyl covers and woven appearing "comfort weave" inserts. When the Interior Decor Group Option (body style 63B) was selected, for example, high-backed, comfort weave buckets were conjured up along with molded door panels lifted from the ultra-plush Mach 1 line and simulated teak wood trim was liberally splashed about the cabin on the dash inserts and three-spoked, rim blow steering wheel. A Rally clock was generally also part of the Decor Group package, but could be deleted.

Other interior options included intermittent wipers, tilt-away steering wheel, a premium sound system consisting of either an AM/FM or AM/FM/cassette radio with dual door-mounted stereo speakers, the interior visibility group (consisting of locking glove box, parking brake warning light, and lighted ignition switch, glove box, ashtray, luggage compartment, and passenger dash cove), and a folding rear seat called the sport deck. Deluxe seat belts and a floor-mounted console rounded out the array of sybaritic comforts available to 1969 Boss 302 buyers.

Although choices were legion in the control cabin of a Boss 302, there were far fewer decisions to be made on the exterior. There were basically three choices: color, sports slats, and rear spoiler. The exciting graphics package cooked up by Larry Shinoda was used to accent one of four available exterior colors in 1969: Wimbledon White, Calypso Coral, Acapulco Blue, and Bright Yellow. Regardless of which hue was ordered, it came outfitted with a low-gloss hood black-out panel, blacked-out head light housing, low-gloss trunk and rear panel, reflective "C" stripes on the body sides, and a front spoiler mounted on the roll pan just below the front bumper. One other body modification that was unique to the Boss 302 line was the elimination of the simulated rear quarter panel air scoops—just as Shinoda had ordered. Flashy Magnum 500 rims and wide oval "shoes" rounded out the Boss' basic exterior package. As mentioned, 1969 buyers could also opt for the unique rear deck spoiler and (or) louvered back light "sport slats."

On the whole, the Boss package was a styling success and traces of Shinoda's genius were soon seen in other manufacturer's cars. *Car and Driver* conducted one of the earliest road tests of the new Boss 302 in June of 1969. The *Car and Driver* testers were generally pleased with the new Boss engine's performance and rated it an ideal compromise for both low-speed acceleration and comfortable expressway cruising. With a wink to the Ford execs who'd limited the Boss' official announced horsepower, the road test offered that the new small block made "at least" the 290 ponies that were claimed for it.

As much as the *Car and Driver* guys liked the free-revving Boss motor, it was the suspension that really got them excited and they said as much. Ride quality was found to be on the stiff side but not unacceptably so. Understeer was reported as greatly reduced over previous iterations of the Mustang line, to the point that the testers were able to drive—rather than plow—through corners. In general they rated the Boss as "the best-handling Ford ever to come out of Dearborn and (possibly) the new standard by which everything from Detroit (is to be) judged." *Sports Car Graphic* offered its first editorial comment on the Boss 302 the same month as *Car and Driver* and featured a studio car on its June cover. Though not quite as hyperbolic about the new suspension and engine package as the *Car and Driver* crew had been, the *Sports Car Graphics* testers still concluded that the Ford's new Boss was easily the equal of the Chevrolet Z-28.

Like all 1969 Mustangs, Boss 302s featured four headlights. Only Boss cars carried this racy black out treatment, however.

52

THREE

The 1970 Boss 302

When the 1969 season concluded, Ford's Trans-Am effort might have been down but it was definitely not out of the picture. Though Ford's two factory backed teams had come into the season powered by entirely new and only recently developed engines and an equally new chassis package, they'd dominated the first races on the circuit and seemed set on a championship winning course until the disastrous wreck at St. Jovite. It's likely then that both the Shelby and Moore teams harbored optimism about the upcoming 1970 season even as the old season wound to a close.

Unfortunately, the Ford racing world was turned upside down in August of 1969 by an off track event almost no one would have predicted the day before it occurred: Bunkie Knudsen was fired. According to a *Car Life* account of Knudsen's dismissal just a handful of days before the 1970 model lines were to be unveiled, "Ford didn't say what went wrong, or what didn't go right. It may have been the parent corporation is too big for one man to control and that Knudsen insisted on controlling it." Later *Car Life* accounts placed the reason for Knudsen's summary dismissal on clashes with Henry Ford II. And a number of those clashes were said to have been about the corporation's racing endeavors. More specifically, it was said that some within the corporation wanted to draw back from Knudsen's aggressively pro-racing stance. Knudsen acolyte, Larry Shinoda saw things more simply; he said at the time, "When you fight an alley fighter and you're not an alley fighter, chances are you're going to lose your tail." In this particular case,

the alley fighter that Knudsen lost his tail to was none other than Lido Iacocca, the same man who'd pushed so hard to establish the Mustang's sports car credentials five years earlier. Though Iacocca had once been an avid backer of factory Mustang racing, he'd apparently lost that enthusiasm by the time he took Knudsen's place at the helm of the Ford Motor Company. Proof of that fact is the withering 75 percent decrease in factory motorsports sponsorship that Iacocca announced (across the board) almost before he'd had a chance to warm the leather of his executive suite chair. NASCAR, NHRA, and SCCA budgets were all to be scaled back dramatically for 1970 and eliminated entirely as soon as possible thereafter. According to Iacocca, Ford's future lay with turgid but thrifty car lines like the Maverick and Pinto—and NOT the Mustang. It was truly the beginning of a long bleak period in Ford performance history. One the corporation did not truly recover from until the mid-1980s, in fact.

Knudsen's departure was followed in short order by Shinoda's (and that of many of the others that Knudsen had brought with him from the General—including Smokey Yunick). Even so, their dismissal did not take place before significant changes had been made in the styling for the 1970 Mustang line. And not before the ground work had been laid for Ford's return to glory on the Trans-Am circuit.

Iacocca's decision to gut the corporate racing budget was not well received. But it was official Ford policy, and so, had to be followed. At first, the 75 percent cut in funding seemed to indicate that just one Ford factory team would carry the Ford banner into battle in 1970. Carroll Shelby's decision to shut down his racing and production car operations following the 1969 model year made implementing Iacocca's cuts somewhat easier but not necessarily painless. Early rumors had Bud Moore downsizing his team to just one car piloted by Parnelli Jones. For a while it looked as

1970 Boss 302s featured restyled front fender extensions and only two headlights. A front spoiler was also part of the new package.

55

Ford engineers were confronted with a rules change for 1970 that limited their team cars to just one four-barrel carburetor. Their response was the Ford Autolite In-line carburetor. The In-line featured four generously proportioned in-line throttle bores. Though it incorporated a variety of stock components lifted from "normal" Autolite carburetors, it was actually a purpose built racing piece. And that's just what the SCCA concluded. As a result, Ford was required to make at least 500 In-line carburetors available to the general public in order to homologate its use on the Trans-Am circuit. SCCA tech inspectors were never quite satisfied that Ford had done so, as a result, the In-line carburetor was never allowed to race.

Thwarted in their hopes to use Ford's new In-line carburetor, Ford racers turned to Bud Moore for help. The ram box intake he helped develop came to be called the Bud Moore mini-plenum. It worked well enough to win the 1970 championship.

though Holman & Moody would step in to fill the gap left by Shelby's departure, and reports to that effect appeared in the motoring press.

Ultimately, the task of winning the Trans-Am title was left to Bud Moore's Spartanburg, South Carolina-based concern and shortly before the first race of the season it was announced that both George Follmer and Parnelli Jones would return as Moore team members.

The sweeping changes in Ford's front office were far from the only differences that awaited Moore, Jones, and Follmer on the 1970 Trans-Am circuit. In fact, during the off season, the sanctioning body had almost completely rewritten the Trans-Am rules book. For one thing, the Automobile Competition Committee of the United States (ACCUS) decided to standardize the pony car rule between the NASCAR GT (Grand American) and SCCA Trans-Am divisions. Though the formula employed by those two series had been close in the past there were sufficient differences to prevent cars from competing on both circuits. The rules book for 1970 also upped

a Trans-Am race car's overall weight to a girthsome 3,400lb wet (or 3,200lb dry). The dual four-barrel carburetors that most teams had employed in 1969 also went the way of the Dodo bird. In their place, the rules book mandated the use of a single four-barrel carburetor, though that fuel mixer was allowed to be a huge 1050cfm Holley Dominator. Another significant change in the engine bay wrought by the rules book was the decision to permit manufacturers to base their Trans-Am cars' race engines on powerplants that had been destroked from their stock displacement. And that made life a lot easier for both Chevrolet and Mopar teams during the 1970 season. Whereas Ford and Chevy had to build special (expensive) street versions of their corporate racing engines in 1969, the new SCCA rule allowed manufacturers to homologate their 1970 race mills with engine packages that displaced far more than five liters. Chevrolet, for example, was able to legalize its racing Z-28s with a fleet of 350ci-powered everyday drivers. Dodge and Plymouth teams took a similar tack by building a sufficient number of 340ci Challenger T/As and AAR 'Cudas. Ford opted to continue production of the Boss 302. One other new-for-1970 tech requirement was the rule limiting transmissions to four forward speeds. Furthermore, any racing modifications made to those transmissions had to be made available to the general public via their local dealerships.

The rule change of greatest overall importance (at least to corporate bean counters, that is) was the new requirement that each manufacturer build a homologation run equal to 2,500 units or 1/250th of the 1969 production (whichever was greater) of any chassis package intended for competition on the Trans-Am circuit. As a result of this new rule Ford was required to build far more then the 1,000 street Bosses it had taken to legalize the line for competition the year before. Under the new formula, Ford was required to crank out at least 6,500 street versions of the Boss 302 before a single racing counterpart could fire an engine in anger.

Interestingly, though chassis production was set at (in the case of Ford cars) 6,500 units, the requirements for specialty engine production were kept far lower. Just about 6,000 units lower to be exact. For 1970, a manufacturer wishing to campaign a special engine or power train package needs only to make 500 versions of that component of engine available to the buying public. And, in fact, that's just what Ford tried to do with a new exotic carburetor that had been devised to satisfy the new SCCA single four-barrel rule. The new fuel mixer, referred to as the Ford "In-line" four barrel was based on a unique four throat alloy casting that featured Weber-like throttle bores. Though outfitted with a wide variety of standard Ford Autolite carb components, the new In-line 4Vs were intended to serve as pure race components. And that's just what the sanctioning body concluded prior to the 1970 season opener. When Ford was unable to convince the Trans-Am officials that the 500 odd In-lines that had been built were actually available over the counter, the carbs were disallowed for use in competition.

Although the rules governing the 1970 Trans-Am season were changed in a great many ways, the cars that Bud Moore campaigned were quite similar to the ones that had seen duty in 1969. As a matter of fact, in some cases they were the same cars that had been raced in the preceding season. Several of Moore's 1970 team cars were newly constructed, or, at least, had been towards the end of the 1969 season. In that

Boss teams faced both new rules and new competition during the 1970 season. Plymouth waded into the fray with a two-car team of fast but unreliable AAR Barracudas driven by Dan Gurney and Swede Savage. Sam Posey headed up a Dodge Challenger effort that nearly snatched the title from Ford in 1970. Roger Penske shifted his racing allegiances to American Motors for 1970, with Mark Donohue behind the wheel of a red, white and blue Penske Javelin. In the photo above, Parnelli Jones (15) and Mark Donohue (6) lead a multi-colored group into a corner. Note the aluminum front spoiler on Jones' Boss. Craft Collection

number were two particular 1969 "M" code chassis that had been fitted with unique rear differentials that featured "free floating" union between the axle housing and the rear leaf springs. Instead of the conventional fixed mounting points, a three-point, heim joint-equipped lower link worked in conjunction with race-proven override traction bars to keep the housing located squarely beneath the chassis. The new setup received mixed reviews from team drivers, with Follmer liking it and Jones being somewhat less enthusiastic. Another suspension change for 1970 was the substitution of lightweight four-piston calipers for the cast iron Lincoln stoppers relied on in 1969. Beyond those few differences, however, the chassis under Bud Moore's 1970 team cars was pretty much identical to his 1969 Bosses. In response to the SCCA's new induction rules, Bud Moore helped design a new intake and carburetor package to top off a race spec Boss engine. The intake featured a cavernous single-plane plenum that was capped off by a single Holley four barrel. Testing on Moore's Spartanburg dyno provided promising figures and ultimately the intake setup came to be called the "Bud Moore mini-plenum" by Ford racers during the 1970 season. It was no doubt called something else by the Brand X drivers who were unable to catch it or the cars it was bolted to during the balance of the competition year.

One other change Bud Moore made to set his team cars off from the pack was to jettison the red, black, and white livery that had been his racing trademark in favor of what came to be referred to as a school bus yellow paint scheme. Other bits and pieces of the new cosmetic package included the long-since homologated (but yet to be raced) rear deck spoiler and a set of 1970 style Boss stripes mounted farther down on the body work than those found on street cars. 1970 team cars, more often than not, relied on feather weight magnesium Minilite rims rather than the crack prone 200S wheels that had predominated in 1969. In addition to the other changes to the Trans-Am series that had taken place during the off season, the complexion of the starting grid was also radically altered by Roger Penske's decision to abandon his long time corporate sponsor, Chevrolet, in favor of a factory backed fleet of AMC Javelins. Donohue once again reprised his role as primary team driver and the red, white, and blue Javelins he campaigned were blindingly fast.

Bud Moore changed racing livery for 1970. School bus yellow (not Grabber orange, by the way) was his hue of choice for 1970. Parnelli Jones made sure the color showed up with regularity on the sports page by winning more than a few events. At season's end, he and teammate George Follmer had recaptured the Trans-Am title for Ford. Craft Collection

Dale Sale's Jones car is virtually the same as when it was raced in 1970. As a result, it is literally a rolling piece of history. When not on the vintage race circuit, it is usually on display at the International Motorsports Hall of Fame in Talladega, Alabama.

The void in the "General's" ranks left by Team Penske's departure from Chevrolet was filled by the arrival of laconic Texan and legendary road racer Jim Hall's arrival. Famed as the innovative creator of the prototype Chaparral racing cars that had literally put the sports car world on wings, Hall's decision to campaign a fleet of refrigerator white Camaros for 1970 gladdened Bow Tie Lover's hearts everywhere.

The other big news along pit row for 1970 was the arrival of not one, but two different Mopar-based Trans-Am efforts. Of greatest interest to the racing cognoscenti was Dan Gurney's decision to field a pair of AAR (named for his All-American Racer team) 'Cudas that were powered by a potent destroked version of the 340 Chrysler engine. Gurney, in addition to overseeing team operations, was also responsible for driving one of the team's two "Gurney Blue" 'Cudas. Long time Gurney protégé Swede Savage got the nod to drive the second AAR 'Cuda for 1970. Sam Posey edged out long-time Mopar partisan Bob Tullius to take factory sponsorship for a third Mopar-based team. In this case, an organization built around a fleet of race spec Challenger T/As. With all of Mo-

The popularity of the vintage race series has brought a number of former Trans-Am cars out of the woodwork in recent years. Jon Vanprooyen recently restored this independently campaigned Boss.

town's major car manufacturers in the fray, the 1970 Trans-Am season promised to be one hell of a contest. It didn't disappoint.

The series opener for 1970 took place at the lovely Laguna Seca circuit located on the Monterey peninsula in California. Things heated up at Monterey well before the first engine was fired along pit road due to a new-found enthusiasm by SCCA officials to enforce the loophole-riddled rules book. Moore's two Mustangs came under fire early on in the process due to the In-line carburetors. Tech inspectors also took issue with the front brake cooling ducts that Moore had routed from the new for 1970 front fender extensions. Both the carburetors and the duct work were ultimately disallowed. The fact that Moore's rivals were also recipients of the tech team's not so tender mercies (Firebird teams lost their Ram Air bonnets and Hall's Camaros were forced to run sans rear spoilers) was probably of little consolation to Moore and his team drivers as the green flag fell. Their collective outlooks had brightened considerably by the end of the race.

Jones and Follmer dominated the field in qualifying, and built on their qualifying advantage during the race to leave Donohue and the rest in their dust to the tune of a half-second a lap. By the time the race was just past halfway, Jones' number 15 Boss had lapped the entire field save for Donohue. Jones maintained that pace until just before the checkered flag when he slowed just enough to permit Follmer to draw close for a two Boss photo op at the finish line.

Lime Rock was the next stop for the Trans-Am circuit in 1970 and once again Jones charged to an early and commanding lead. While his rivals and even teammate Follmer fell prey to a variety of mechanical woes, Jones' Boss ran flawlessly—and fast. By the first pit stop, Parnelli's lead was so great he was able to get in and out of the pits without allowing the second-place car to pass. By the end of the two-and-a-half-hour contest, he was a full lap ahead of the rest of the field and coasted home an easy winner even though his brakes and one cylinder were almost gone.

Ford engineer, Ed Hinchcliff, built his own independent Boss 302 Trans-Am car with input and parts from Kar Kraft engineering in 1970. He drove the car on the 1970 Trans-Am circuit as an independent. Terry Bookheimer found the car abandoned in a salvage yard several years back and returned it to its former glory. It is one of the fastest, most attractive cars in the vintage ranks today.

Things didn't get any better for Donohue, Hall, or the rest of the pack at Loudon, Hew Hampshire in May, even though Swede Savage's AAR car was fastest in qualifying and in the early stages of the race. When Savage's clutch failed at the midway point, Jones assumed the lead. When a freak accident ripped Jones' hood completely off the car, it was Follmer's turn to lead. He maintained that position until the end of the race. Three races into the season Ford had built up a commanding lead in the points race and all was well.

Ford's fortunes seemed to take a turn for the worse during qualifying for the June 7th Trans-Am race at Mid-Ohio. Moore's teams cars were both off the pace in practice and both Mark Donohue and Ed Leslie capitalized on that fact by claiming the front row of the starting grid. The AMC team leader had made a preseason promise to win at least seven races for his corporate sponsor and was no doubt encouraged by team driver Donohue's performance. But there's a big difference between optimism and winning. And that's just what Messrs. Jones and Follmer proved to Penske as soon as the green flag unleashed the field. Though denied use of the new Autolite In-line carburetor a second time by the sanctioning body, Jones and Follmer turned the Mid-Ohio 180 into a Ford benefit. Jones vaulted into the lead on the first lap and held it until his first pit stop. Jones quickly recaptured the top spot on the track after fueling and by race's end had built up a 68sec lead over all but Follmer's second-place Boss. The race boiled down to a battle between Jones and Follmer. For a time both driver's pit crews (and Ford executives in attendance trackside) despaired that Jones and Follmer's fender-rubbing might ultimately knock both of them out of the event. Especially when Follmer completely ignored his pit board-flashed instructions to back off and let Jones take the flag unscathed. Jones did prevail, and Follmer was heard wondering out loud just how someone was supposed to pass Parnelli's exceedingly wide Boss

Here's an early publicity photo of a prototype 1970 Boss. Note the car's non-regular production argent, tall center cap-equipped wheels, and black grille treatment.

302. The obvious answer was you weren't supposed to pass Parnelli.

Ford's domination for the Trans-Am series had been so complete during the first quarter of the season that Penske's corporate sponsors were on the brink of pulling the funding. Team Javelin's fortunes would have to be reversed dramatically if those red, white, and blue AMCs were to continue on the Trans-Am trail. Unfortunately for Ford, that's just what occurred at the next stop on the circuit at Bridgehampton. Perhaps it was inevitable that Penske's immaculate preparation and Donohue's inspired driving would produce a win. That first AMC win came in June at the Marlboro 200. Pre-race events were dominated by another rejection of Ford's bid to field the In-line four-barrel carburetor. Parnelli Jones, for one, was disgusted by the actions of the sanctioning officials and remarked to the press that, "They just don't want Ford to win this series!" While SCCA's malice toward Ford remains unknown, the SCCA certainly would have preferred close racing to a season of Ford dominance.

Race day at Bridgehampton dawned rainy and worsened throughout the race. Donohue put his wet track driving skills to good use by passing pole sitter Savage and both Ford cars to take the lead. SCCA pit workers did their part to help Donohue hold on to that advantage by penalizing Follmer during a pit stop for too many men over the wall. Jones' sealed Ford's fate when he overshot his stall in the blinding rain and then did a bit of off-track "agricultural racing" when

Medium Lime Metallic looked good when decked out in Boss stripes and spoilers. Note the black-centered Magnums that were optional in 1970.

he blew a tire. The final tally was Javelin first followed by Bosses second and third, which solidified the Ford points stranglehold. Upon climbing out of his car, Jones again offered his observation that the sanctioning body, "didn't want Ford to win this series." Donohue's win at the Bridge was the first in a string of disappointments for Moore's Boss 302 organization.

The next race on the circuit at Donnybrooke produced Chevrolet's first Trans-Am win of the season. And the July 19th event at Road America in Wisconsin saw the Penske Donohue Javelin team visit victory lane for a second time in three races. That particular loss was made all the more painful by the mid-race shunt that took the life of former Shelby team driver, Jerry Titus. Ford fortunes did not improve in August when the series returned to St. Jovite. Alarmed by Penske's points race encroachment on the Boss teams' lead in the championship, Moore prepared a Boss for A.J. Foyt to campaign in Canada. Jones set fast lap during qualifying and Follmer started the race just a row behind Parnelli's pole winning grid position. The early laps of the event were a Javelin "sandwich" affair as Jones and Follmer ran first and third to Donohue's second. But an early pit stop strategy that team Penske had been practicing for several events ultimately produced a lead for Donohue's red, white, and blue racer that Follmer, who eventually came home second, couldn't overcome.

The largest field of entrants for the whole season showed up in the paddock area of the Watkins Glen circuit for the ninth event on the Trans-Am schedule. Qualifying was once again dominated by Jones and Donohue and their two cars sat on the pole. Jones led for the first third of the event but was forced to surrender that advantage for a fueling stop. Rain visited the fabled New York circuit about the same time as the first round of pit stops, further muddying the racing order on the track. When the field had switched to rain tires and returned to the track, Vic Elford took

The biggest underhood change for 1970 was the addition of Ram Air as an option. When ordered, that option consisted of a cast alloy scoop bolted to a modified air cleaner base. With the hood closed, the new assembly jutted through a trim-equipped cut out and bobbed menacingly in response to engine movements.

Big Boss Man; Walter "Bud" Moore

It's likely that Bud Moore didn't consciously set out to be one of the most successful road racers of all time, but that's just what happened. Like many in his generation, Moore responded to Uncle Sam's call during World War Two and soon found himself slogging ashore on a beach in Normandy. After making it across the killing zone on that French beachhead, Moore followed old "blood and guts" Patton across the continent. In six months of fighting he saw action in five major battles, received two purple hearts, two bronze stars, and was promoted to the rank of Sergeant.

At the close of hostilities, Moore returned to his Spartanburg, South Carolina home and began to dabble in the used car business. In time he found himself wrenching on cars for many of the local modified drivers. Though not a driver himself, Moore had an uncanny ability to make a competition car go fast. Soon, his skill as a mechanic brought him to the attention of drivers and crews on the fledgling NASCAR circuit and, by 1957, he'd become the chief mechanic for Buck Baker's factory Chevrolet team. One of Moore's earliest road course wins came when a car he'd prepared for Baker finished first at Watkins Glen that season.

After working as crew chief for Speedy Thompson and Jack Smith, Moore decided to start his own Grand National team in 1961. Moore's first team driver was Little Joe Weatherly, who was alternately called the "Clown Prince of Stock Car Racing" and the best driver on the circuit. Both titles were appropriate. Weatherly and Moore were an immediate success and together they won the Grand National driving championship in 1962. Their string of wins continued in 1963 and Weatherly cinched

Bud Moore's first team car was the second race-spec chassis completed by Kar Kraft. Like Shelby team car number one, Moore's first Boss also started life as a 428 Cobra Jet-powered car (serial number 9F02R112074). That car was the first to do battle and it made the Boss 302's competition debut at Daytona in February 1969 at the Citrus 250. The fared over headlight buckets and NASCAR-style side fuel fill were removed when the car became George Follmer's primary car on the SCCA circuit later in the year. Craft Collection

his second straight Grand National title with a road course win at Riverside International Raceway in California. Tragically, Little Joe lost his life at the very next Riverside race when his Bud Moore-prepped Mercury went wide in RIR's turn six and hit the wall. Moore signed Billy Wade to fill Weatherly's seat for the balance of the year and Wade notched two more road course wins for the team (at Bridgehampton and Watkins Glen) along with several oval track triumphs before he also lost his life on the track, in this case during a tire test.

Moore carried on as a team owner in the Grand National ranks until 1967 when Lincoln Mercury called him away to field a team of Trans-Am Cougars on the SCCA circuit. Moore tapped Dan Gurney, Parnelli Jones, and Ed Leslie as team drivers and prepared a fleet of red and silver Cougars for them to drive. Moore's skill at setting up a road race chassis quickly became evident and soon his Cougars were giving Ford's Shelby American Mustang team fits. Fact of the matter is, Moore's Cougars almost took the title away from Shelby's Terlingua team. And that gave Ford execs such a start they decided that one season of factory-backed Cougar racing had been quite enough. All told, Moore's Cougars won four races in 1967 and sat on the pole at five. At season's end, Moore's Cougars were a mere two points behind Shelby's title-winning effort in the points championship.

Parnelli still has fond memories of his first pairing with the tall quiet man from Spartanburg. "I got along with him very well," Jones recalls. "We were on the same path together. The thing I liked was if I wanted something changed, boy, they (Moore and the team) went right after it. I enjoyed it. Bud's a super guy." Those good feelings would ultimately lead to Jones and Moore teaming up on the 1969 Trans-Am circuit.

But, back to 1968, Moore's immediate success in Trans-Am racing (read: he beat the tar out of Shelby's Ford team Mustangs) led to the withdrawal of funding for that effort in 1968. So Moore took his team Cougars home to NASCAR country where he campaigned them in the Grand American division (NASCAR's version of Trans-Am) with the help of journeyman racer Dewayne "Tiny" Lund. The end result was a Grand American national championship.

Shelby's Trans-Am team lost the Trans-Am title to Chevrolet in 1968, so perhaps, it was only natural for Ford to recall Moore from NASCAR to head up a Boss 302 Mustang team effort for 1969. Parnelli Jones was the first driver Moore signed after accepting the assignment. George Follmer also got the nod to pilot one of Moore's red, black, and white Mustangs for the 1969 season. Moore began work on the 1969 SCCA season in the Fall of 1968, when Jacque Passino called him to Dearborn for a conference. As Moore recently recalled, "They wanted me to take over the Trans-Am series in 1969 and run the Mustangs. They [Ford] were already testing on the West Coast at Riverside. I left Detroit to fly out to Riverside to start overseeing the testing program." But Moore was not pleased with what he found out on the West Coast. "They had about 15 engineers. Nobody knew what was going on. I called back to the Ford people in

[Michigan] and told 'em, if I was going to run this series, we're gonna' have to make some changes." Moore ultimately worked with a pared down engineering staff that consisted of Lee Morse (who is currently SVO's boss), Lee Dykstra, and Harold Grasse. Though history records that Moore's team was the most successful of Ford's factory efforts (Jones and Follmer scored three wins to the Shelby team's one), it wasn't enough to prevent Penske's Chevrolet team from reclaiming the SCCA title.

Things were different in 1970. When Ford pared its racing efforts, Moore's Trans-Am team was the only factory effort to keep its funding. Jones and Follmer returned as team drivers and together they campaigned a duo of yellow Boss 302s on the 1970 circuit. As in 1969, success came early for the Moore operation. Jones tasted victory first with wins at Laguna and Lime Rock. Follmer visited victory lane at Bryar and by year's end Moore's Bosses had scored six wins in eleven races and the season championship.

1970 was the final year of Ford factory backing for motorsports competition. Even so, Moore opted to soldier on in SCCA competition with the Boss cars he had left over from the 1970 season. Though George Follmer scored two more Boss 302 wins for the team, the lack of factory backing was telling.

When Moore returned to the NASCAR ranks the following season it was a much different series than the one he'd left. Concerned by the speeds that the big-block engines of the day were generating, NASCAR officials were desperately searching for ways to slow things down. Though restrictor plates had been introduced to that end in 1970, speeds had not been trimmed enough and smaller plates had been introduced. Moore was quick to notice that NASCAR's restrictor plate rules looked with more favor upon small block engines than mountain motor Boss 429s—if anyone chose to run one.

And that's just what Bud Moore did. His three year's experience with Ford's cant-valve headed engine families prepared him for that task. The white number 15 Torinos he built for drivers like Bobby Isaac, Buddy Baker, and Darrell Waltrip were innovative and successful. Ultimately, they led the way to NASCAR's small-block power future. Bud Moore is still campaigning evolutions of those same small-block engines on the NASCAR circuit.

Bud Moore was always an innovative team owner. He was one of the first to "downsize" in 1966. The nimble little Mercury Comets he built that year were the first Ford intermediate Grand National stock cars on the circuit. When Moore turned to campaigning Boss 302s, he and drivers Parnelli Jones and George Follmer won the 1970 SCCA Trans-Am title.

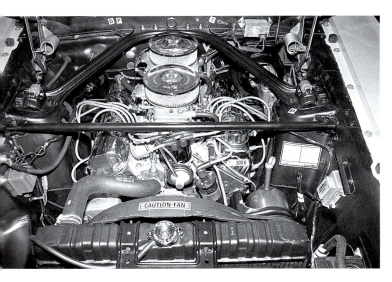

The aftermarket industry began to gear up in response to Ford's new Trans-Am small-block engine by 1970. This particular Boss engine is fitted with a "Shelby" lettered dual four-barrel intake that was available over the counter at many Ford dealerships. Note the intermittent wiper control box on the left firewall—a very rare 1970 option. Also note the chrome valve covers that early 1970 Boss Mustangs carried. Shelby's aftermarket high-performance parts company also cast up a handful of "Shelby" lettered Weber intake manifolds for use atop Boss 302 motors.

command in a Jim Hall-prepped Camaro Z-28. Donohue took second place when he squeezed by Follmer's momentarily fuel-starved Boss on the last lap and Jones took fourth behind Follmer. With just two events left on the circuit, the points race had narrowed to 60 for Ford and 49 for AMC.

The tenth race of the 1970 Trans-Am season was held in Kent, Washington. During qualifying Jones was again the fastest—twice. After setting the fastest lap in his own number 15 Boss, Jones went back out in the team mule and shaved two-tenths off his best previous time. Follmer started from the second row. Jones took the early lead and was not foiled by either Penske pit strategy or mid-race rain. The hard-charging Jones eventually built up a nearly 20sec lead over Donohue and held on to take both the race and the season championship. Ford had finally recaptured the Holy Grail of pony car racing—the SCCA Trans-American championship!

Although the season was effectively over at Washington, one race remained, the Mission Bell 200 at Riverside International Raceway, California . Not one to rest on his laurels, Jones blistered the RIR road course during qualifying with a record 103.67mph lap. Follmer was also hot during pole testing and his performance earned a front row starting berth for the number 16 Bud Moore-prepped Boss. Jones and Follmer capitalized on their qualifying dominance

A variety of Grabber colors became available in 1970. Grabber blue looked particularly fetching when accented by the Boss black out and tape package.

during the race proper and kept both of their Bosses out front for most of the event. Jones' performance in winning the race that day was made all the more impressive by the mid-race, door-crunching punt into the desert from a wandering back marker. By the time Parnelli had regained his purchase on the asphalt, his Boss was back in ninth place. In typical fashion, Jones tore into the circuit, setting fast lap of the day on his return to the lead and the win. The final points standings for the 1970 Trans-Am season were Ford-72, AMC-59, Chevrolet-40, Dodge-18, Plymouth-15 and Pontiac-0. As things turned out, Jones' championship was the last major title to be won by a Ford motorsport team (in SCCA, NASCAR, USAC or IMSA) for nearly two decades. It was the end of a memorable era.

On November 20, 1970, Lee Iacocca, still convinced that Ford Pintos and Mavericks were the wave of the future, pulled the plug on all factory sponsored motorsports competition. The fat lady had truly sung on Ford's Total Performance era. Iacocca went on to preside over the tape package "performance" years of the seventies before he moved on to Chrysler.

Bud Moore soldiered on as an independent in the 1971 Trans-Am ranks and again campaigned 1970 Boss 302s (though at one point he'd planned to field 1971 "big-bodied" Mustangs). Parnelli Jones and George Follmer again began the season for Moore, but the team's decreased funding ultimately made Follmer the primary driver. On occasion, Peter Gregg also drove one of Moore's Bosses. The lack of financial depth also made appearances on the 11-stop 1971 tour more than a little problematic and Moore was not able to make all of the events.

With Penske's Javelin effort the only remaining team enjoying factory backing, it was likely that he and Donohue would dominate the year and secure their third Trans-Am title. Although Follmer was able to score two more Boss 302 Trans-Am wins (at Bryar and Mid-Ohio), the season ended in a championship for team Penske. Bud Moore returned to the NASCAR ranks in 1972 where he applied his small-block Ford expertise

The 1970 Dark Ivy Green Metallic prototype also featured a non-regular production black gas cap and a set of sports slats that did not have external hinges. That car also featured a garish vermilion interior. Craft Collection

to the development of 351C-powered Torinos. He never again ventured into the SCCA road course ranks. But even so, the victories scored by his Boss 302 teams in 1969 and 1970 are far from forgotten.

The 1970 Boss 302 on the Street

The SCCA's decision to increase homologation requirements for the 1970 season had a direct and dramatic effect on the street Boss needed to "legalize" Bud Moore's race spec Mustangs. Ford's SCCA obligation was set at 6,500 Boss 302s, a figure far in excess of the River Rouge's capacity (in addition to its pre-existing "white bread" Mustang and Cougar production obligations). As a result, Ford execs opted to devote a portion of their Metuchen, New Jersey plant's Mustang capacity to the construction of Boss 302s. This decision to split Boss production ultimately resulted in the construction of more than 7,000 street legal Boss 302s (7,013 to be exact). Though this production scheme produced production line variations specific to each plant, the Bosses built at Metuchen and in Dearborn were just about identical in all major ways. And according to automotive writers of the day, they were the best all-around Mustangs ever built.

The 1970 Boss 302 Mustangs were based on the 1969 fastback roof lined unitbody. Even so, Ford stylists were able to achieve a fresh and distinctive look for the new model year with the help of a handful of sheet metal changes. Major players in the 1970 Boss' restyling were revised front fender extensions that carried just two head lamps. Whereas 1969 Mustangs of all flavors had come factory equipped with separate high and low beam lamps, styling cues for 1970 collapsed those functions into just two sealed beam units. The resulting spare space at the bow was devoted to a quartet of simulated cooling vents mounted outboard of the head lamps. Though blocked off for street use, Bud Moore's Boss cars made full use of the slot's cooling potential on the track—until SCCA officials declared his clever modification illegal. Farther aft, the concave back panel common to all 1969 Mustangs was tossed in favor of a flat replacement that featured rectangular pot metal light housings instead of the three vertical (per side) tail lamps that had been protruded from a 1969's sheet metal.

Ford also offered just plain "Red" as a standard Boss color in 1970. Boss stripes for 1970 began near the peak of the hood and flowed back over the flanks of the car to the rear bumper. As in 1969, they were cut from reflectorized 3M tape.

In between the two extremes of a 1970 Mustang's silhouette were traces of Larry Shinoda's styling legacy. Gone, for example, were the hokey "C" pillar medallions found on all non-Boss 302 Mustangs in 1969. And the ersatz side scoops that Shinoda had found equally offensive were also gone for 1970. Beyond the Shinoda-inspired changes that impacted the whole Mustang line for 1970, the new Boss 302 cosmetic package also reflected his short but significant term in Ford's styling studio. Take for example the body length, reflectorized tape stripes that graced every 1970 Boss car's flanks. Though starting at the hood's edge and then sinuously flowing back over the front fenders on their way rearward, a 1970 Boss's stripes still owed their origin to the innovative tape adornment first penned by Mr. Shinoda in 1969. Ditto for the standard front and optional rear spoilers that played a significant role in 1970 Boss 302 identification. The racy looking rear window "sports slats" that Shinoda had bolted to his very own 1969 Boss prototype were again optional. Other bits and pieces of the new Boss package for 1970 included a blacked-out upper deck lid, a matching (but narrower than in 1969) black-out panel on the hood, an equally ebony rear body panel, and a reflectorized pin stripe accent on the tail light panel.

1970 Boss 302 Road Test Results

Car and Driver, February 1970
- zero to 40 ....3.3sec
- 60 ....6.5sec
- 80 ....11.1sec
- 100 ....17.0sec
- ¼ mile 14.93sec@93.45mph

Hot Rod, January 1970
- ¼ mile 14.62sec@97.50mph

Super Stock, January 1970
- ¼ mile 14.15sec@100.00mph

Boss buyers' color choices were significantly expanded for 1970. Whereas just four hues were available in 1969, no less than 13 different colors (excluding special order) were available for 1970. In that number were a trio of dazzling choices referred to as Grabber yellow, Grabber blue and Grabber orange. So bright were those new tints (and, of course the returning Calypso Coral scheme) when viewed in direct sunlight, that non-Ray Ban equipped viewing was decid-

Boss 302 engines were available with and without Ram Air induction for 1970. Base engines like this one featured a chrome-plated air cleaner lid and a base that had been fitted with a vacuum diaphragm governed pop-off valve. Note the rev limiter mounted on the car's inner fender apron.

69

light, that non-Ray Ban equipped viewing was decidedly risky. Boss colors also included a flashy non-metallic red that was nearly as eye catching. Those in search of Boss performance less likely to attract attention had at their disposal somewhat more sedate colors like light ivey yellow, white, pastel blue, medium blue metallic, bright blue metallic, medium lime metallic, and bright gold metallic. Special order paint colors were also an option for those 1970 buyers willing to pay the extra "freight" for those choices. All in all, 1970 was a colorful year for Boss 302 Mustangs—both on and off of the track.

In addition to the styling changes, 1970 Boss 302s also differed mechanically from their 1969 predecessors in some fairly significant ways. Perhaps the most important of these was the decision to downsize intake valve size for the new model year in an attempt to improve low rpm performance. As first introduced, Boss 302 head castings carried flapjack-sized, 2.24in intake valves. The large valves made the engine a screamer at high rpm, but produced bucking and bogging when punched at low revs. For 1970, the intake valves were slightly downsized to a more manageable 2.19in that made the 1970 Boss more user friendly.

The 1970 Boss engines were also equipped with non-cross-drilled forged steel cranks. Though margin-

Bright Gold Metallic was one of the rarer Boss 302 hues for 1970.

ally less suited for high rpm lubrication hollow journaled 1969 shafts, a solid journal 1970 crank was far cheaper to produce and probably a bit sturdier, to boot. Beyond those few internal changes and the shift to "DO" casting numbers that reflected the new model year, 1970 Boss engines were pretty much identical to their 1969 predecessors.

The same can be said for the engine's componentry, too. The same free-flowing, cast iron exhaust manifolds, manual choke 780cfm Holley, and dual plane 4V intake were used in 1970. What was new for 1970 was the option of a Ram Air induction system complete with a shaker hood scoop. Checking off the right box on a Boss' order blank produced a vacuum diaphragm equipped air cleaner assembly that was topped off with a cast metal scoop. A scoop that, by design, was so far above the engine that a hole was needed in the hood for clearance. With the bonnet down and the engine at idle, the scoop bobbed menacingly to and fro in response to engine vibrations. In actual operation at full throttle settings, the assembly's vacuum diaphragm pulled open an access door that allowed cooler, denser air direct access to the Holley's four throttle bores. The new induction setup was definitely Boss and it also helped to produce extra ponies.

As in 1969, when a buyer ordered up the optional 4.30:1 ring and pinion, he was simultaneously outfitting his Boss with an accessory oil cooler mounted to the radiator support. The cooler was plumbed to the engine via an alloy adapter mounted between the filter and block and a set of high-pressure oil lines. Though never designated as such on the Ford order form, this combination came to be called the Drag Pack option and is considered rare and highly desirable by Boss partisans. Another underhood Boss addition for 1970 were racy cast alloy valve covers. Though originally announced for 1969 iterations of the H.O. 302, alloy valve covers didn't show up until shortly into the 1970 model year (and after a number of 1970 Bosses had been built with stamped steel chrome plated covers slightly different than the units used in 1969). The 1970 Boss engines also used a different smog pump location and a revised pulley system. The water pump inlet was shifted from the right to the left side between the two model years. As in 1969, 1970 Boss cars came factory equipped with a rev limiter.

As in 1969, the only transmission available was the tried and true "Top Loader" four speed. Available in standard wide-ratio or optional close-ratio versions, a 1970 Boss Top Loader was governed by a new Hurst lever and "H" pattern handle that worked in concert with a set of somewhat less than desirable Ford rods. One other new addition for 1970 was the thief-defeating transmission lock that kept the lever in reverse when the ignition was locked.

A 1970 Boss car's suspension package was far more evolutionary than revolutionary. As in 1969, beefy, large diameter spindles served as the suspension's cen-

In tacit recognition of the fact that valve sizes had been too large in 1969, Ford engineers reduced intake diameter to a still generous 2.19in for 1970. Other critical head dimensions remained identical. Beefy four-bolt mains were still used in 1970. The crankshaft was once again forged steel but was no longer cross-drilled.

tral focus at the bow. Stiff coil springs that were a part of the "competition suspension package" worked in concert with those spindles and a quartet of unequal length control arms to keep a Boss' chin spoiler out of the dirt. And a $^{15}\!/_{16}$in sway bar helped a set of Gabriel shocks defeat body lean in the corners. The same, single-piston, floating caliper, 11.3in diameter discs used in 1969 were employed in 1970 to scrub a Boss' terminal velocity.

A 31-spline equipped, nodular differential was used to bring up a 1970 Boss' rear. And it was held

Brightly hued, plaid cloth inserts were also available in 1970. This particular Boss is equipped with a deluxe "CF" code Ginger and Ginger stripe cloth control cabin.

71

Sybarites could order up plush, Mach 1-style interiors for their 1970 Boss 302s. Simulated teak accents and padded door panels were part of that treatment.

gered shocks. New for 1970 was the ½in rear sway bar that Matt Donner had originally prescribed for the Boss line during research and development of the car's suspension package. Ten-inch drums and another duo of Gabriels finished out the rear underpinnings. Unlike their 1969 stable mates, 1970 Boss 302s did not all roll on flashy Magnum 500 rims. Instead, the base wheel package consisted of stamped steel 15x7 rims dressed up with trim rings and dog dish-style hub caps. A 15in version of the Mach 1 line's "snow-flake" simulated mag wheel hub cap was optional for the Boss line in 1970. The chrome-plated Magnum 500 steel rims were also optional. Regardless of which rim package was selected, a 1970 Boss rolled around under rerolled front fenders on a quartet of F60 wide oval tires.

Ford improved creature comforts by expanding 1970 interior combinations. Gone for good were the standard low-back bucket seats. The replacements were a set of cushy, high-back bucket seats that could be covered in one of 18 different vinyl or cloth

This Grabber yellow car rolls on the base "dog dish" hub cap and trim ring package available in 1970.

72

schemes. Black vinyl was by far the most common choice, but Boss buyers could choose from a rainbow of combinations that were color keyed to the exterior hue. Also new for 1970 was the addition of bucket seats with snazzy, plaid cloth inserts. Checking off the decor interior box on the order sheet called up the same simulated teak wood door and dash inserts that were devised for the Mach 1 line and turned a normally spare Boss car cockpit into a den of earthly pleasures. Except for air conditioning, that is. As in 1969, the only way to cool off in a Boss car was to open the windows and hit the gas.

In addition to choosing between other interior options like a floor length console, a fold down rear "sport deck," an in-dash tach, and the AM/FM sound system, 1970 buyers could order up a tilt (as opposed to the 1969 style tilt-away) steering column, a rim blow deluxe steering wheel, power steering, or intermittent wipers. In fact, with a bit of thought and deliberation, a 1970 Boss buyer could just about create a road race-oriented Mach 1. Interestingly, it appears that most did not. The majority of Boss 302s that remain seem to have rolled off the line with a minimum of interior options.

The Press Speaks

So how did the motoring press react to the new and more sophisticated 1970 Boss? They loved it. *Hot Rod* magazine called the car "A Boss to Like" in their January 1970 road test, and raved about both their test Boss' handling and acceleration (14.6@97.50mph in the quarter). In fact, editor Steve Kelly opined that, "Ford has done their best possible job in coming up with a perfectly suited muscle car that fits street and racing conditions without great amounts of change being required for the transition." And that was heady praise indeed from a publication primarily devoted at the time to extolling the virtues of the small block Chevrolet engine.

Super Stock magazine also put a new Boss through its paces in January of 1970 and discovered that with slicks and a handful of easy modifications, 13.43@103.23 ETs were just a stab of the throttle away. Road tester Jim McGraw was gushy about a Boss' road holding, too and stated that "the Boss 302 will handle just about any kind of road you can get it to with aplomb."

One month later, the irreverent crew at *Car and Driver* hired Sam Posey to test and compare a Boss

In stark contrast to 1969 when color selection was limited, 1970 Boss 302s were available in a rainbow of regular production and special order hues. Among the rarest of those was Pastel Blue.

The 1970 Boss 302 was the end of the line for that model and, some say, the ultimate Mustang.

Silent proof that Ford planned to continue the Boss 302 line for 1971 can be found in the various 1971 coded Boss 302 components that were eventually built. Take this "D1" (i.e. 1971) engine block, for example. Such 1971 mechanical components became service items when the decision was made to drop Boss 302s in 1971. Craft Collection

302, a 289 Cobra, a 340 Duster, and a 454 Chevrolet. An eclectic gathering, that. The venue was Posey's home track (and site of his 1969 Shelby Team Boss 302 win), Lime Rock. Though Posey carped about his test Boss' manual steering, he still had favorable things to say about the car's road handling and refinement. The Boss was also given high marks in the braking department. As expected, the Cobra was the fastest around the course at Lime Rock, but Posey's laps in the Boss were impressively close—less than three miles per hour slower, in fact.

Some initial steps were taken towards the production of a 1971 version of the Boss 302 (some 1971 parts numbered Boss 302 pieces have turned up over the years), but that version of the car died on the vine. In its place, Ford introduced the Boss 351. And that's a story we'll deal with a few more pages down the line. Today, 1969 and 1970 Boss 302s are revered by Mustang enthusiasts and fondly remembered by those who once owned them. Of the 8,641 street going Bosses that were ultimately built, fewer than 2,000 have been accounted for, and fewer still have been restored to their full former glory.

Production Differences Between Dearborn and Metuchen Bosses

Boss 302 production was expanded to the Metuchen, New Jersey assembly plant in 1970 in order to meet the higher homologation requirements the SCCA set for that year. The River Rouge plant in Dearborn also continued to build Boss 302s for 1970. Though the Trans-Am ponies that rolled off of those two lines were identical in most major ways, there were a few production differences between the "F" code (the second character of a 1970 Ford VIN number identified a car's production site) Dearborn and "T" code Metuchen cars.

Dearborn	Metuchen
Build tag mounted on the right side of the radiator core support	Build tag located on the upper right fender apron, near the hood hinge
Tag smooth edged	Tag wavy edged or smooth
Behind grille surface of headlight sprayed (crudely) black	Behind grille surface of headlight buckets left body color
Underneath trunk, hinges, and majority of interior of trunk painted black	Under trunk remains body color
Trunk rod clip mounted low	Trunk rod clip mounted high
Rear stripes mounted low	Rear stripes mounted high
Engine decal staggered, one front, one rear	Engine decal not staggered both at front
Choke knob mounted closer to dash centerline	Choke knob mounted farther from dash centerline
Trunk/end cap pin stripe mounted high	Trunk/end cap pin stripe mounted low
Engine box wiring harness runs to blower motor then shock tower	Engine box wiring harness runs beneath right side firewall brace

Dearborn cars carried their tail panel pin stripes higher and their rear quarter panel Boss stripes lower than Metuchen cars.

The rear panel black out treatment on Metuchen cars was tapered at the top but stepped at the bottom (as shown) in 1970. Dearborn cars featured black out paint treatment that was tapered both top and bottom.

FOUR

The Boss 429

You might say that Ford's Boss 429 was born at the Daytona 500 in 1964. For that's the day that Ford-backed NASCAR racers first tasted defeat at the hands of a gangly fellow named Petty from North Carolina and his 426 Hemi-powered Belvedere. As you might imagine, that win and Petty's subsequent domination of the Grand National ranks that season sat none too well with the folks at Ford. Adding insult to injury was the fact that Chrysler's Hemi was anything but stock—even though Big Bill France made a big show about only allowing "stock" American automobiles into competition in his NASCAR Grand National division. Fact of the matter was, when Petty first visited victory lane in his Blue number 43 Plymouth, the build date for the first regular production Mopar Hemi car was still nearly two full years in the future. The Hemi engine that Petty used to bludgeon his Ford rivals was nothing less than a purpose-built, full racing engine. And NASCAR officials knew it. Yet, it was allowed to run "for the good of racing." Ford was in high dudgeon about it.

Official complaints about the engine's homologation status were flying south to France's Daytona Beach headquarters even before the starting grid for the 500 was summoned onto the track. And those protests grew louder with each passing day (and additional Hemi victory) that followed. Unfortunately for Ford racers in 1964, France chose not to respond to those complaints during the season. As a result, Petty and his fellow Hemi drivers romped through the competition and Petty was the Grand National driving champion by season's end.

The very first Boss 429 built rolled off of Kar Kraft's Brighton assembly line on January 15, 1969. Craft Collection

Petty's less-than-secret weapon was the all-new engine under his hood. Evolved from the Chrysler wedge motors that had been campaigned the preceding season, the 426ci motor at Daytona was topped by a set of radically configured, hemispherically chambered cylinder heads. Unlike the typical "wedge" motor, with its in-line valves, unequal length intake and exhaust passages, and wedge-shaped combustion chambers, the new Hemi was cast with equal-length intake and exhaust passages that centered on dome-shaped combustion chambers. The new Hemi heads flowed better than the Holland tunnel. Better yet, their basic configuration helped suppress detonation—which in turn permitted the use of higher compression ratios. A conventional wedge-headed motor just didn't stand a chance when compared to a Hemi-headed engine. And Ford knew it.

And so, even while continuing its loud and vociferous protestations about Chrysler's Hemi, Ford engineers began their own development of a Ford-built Hemi-headed racing engine. As NASCAR fans might recall, Big Bill France reversed his legalization of the Chrysler Hemi for 1965, and as a result, Dodge and Mopar teams boycotted the series. With their chief rivals on the sidelines, Ford teams had a field day. That success took some of the urgency out of Ford's in-house Hemi work, but it was still carried on. By 1966, that work had produced a special Hemi-headed single overhead cam version of the corporate 427 racing engine that cranked out more than 650hp in peak tune.

Ford execs called the new engine the Cammer (due to its then innovative cam train) and petitioned NASCAR officials to deem it legal for Grand National competition. Unfortunately, Ford's new Hemi was no more "stock" than the Chrysler Hemi had been in 1964. Though Ford made a show of supposedly building a handful of "Regular Production" Cammer-pow-

77

Many of NASCAR's most famous mechanics and drivers campaigned Boss 429-powered Grand National cars during the 1969 through 1974 seasons. Here Banjo Matthews (glasses) runs a race Boss on a engine stand at his Asheville, North Carolina shop.

ered Galaxies (one such car was given to astronaut Scott Carpenter with much hoopla and fanfare), France was not deceived. As a result, the Cammer Hemi was disallowed for competition in 1966. Henry Ford II yanked his Ford teams out of the Grand National series in spite, thus handing Dodge and Plymouth drivers the season on a silver platter. Making matters worse for Ford partisans was the fact that Chrysler had decided to up the horsepower ante by (belatedly) building a sizable number of regular production Hemi-powered street cars. As a result, not only did Chrysler racers dominate the 1966 season, but they did it with fully homologated Hemi engines.

France induced Ford to return to the NASCAR fold by bending his own rules book a bit (to permit the use of decidedly non-production Tunnel Port 427 wedge head castings) but he would not relent on the Cammer issue. Had he done so, Chrysler had threatened to unleash a dual overhead cam version of its Hemi that would have escalated the horsepower beyond imagination. And so Ford executives and racers soldiered on in NASCAR competition still jealous of the Chrysler but forced to race without one of their own. That would soon change.

Ford engine and foundry engineers were nothing if not prolific in their creation of new engine families

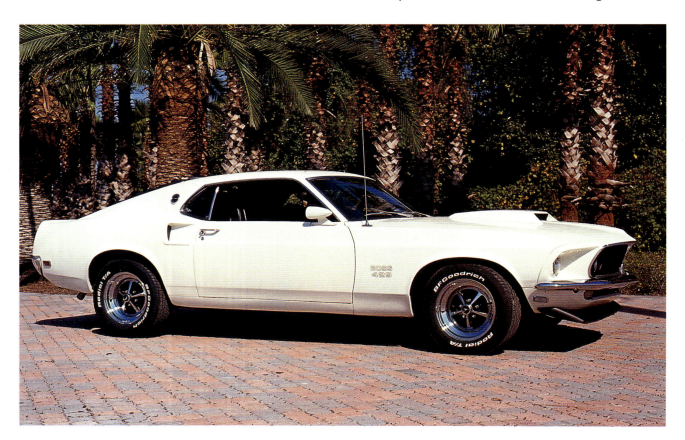

Wimbledon White Boss 429s looked particularly "white bread" from a distance. Until, that is, their low profile B-O-S-S 4-2-9 fender badges came into view. Hushed respect soon followed that flash of recognition.

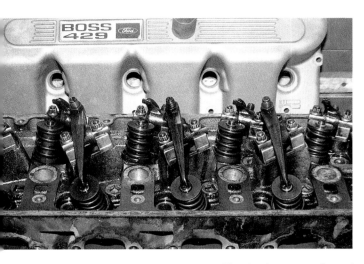

Boss 429 engines featured "twisted" valve layouts and novel rocker arms. Early engines used magnesium valve covers but most Bosses came with aluminum covers.

Intake ports were generously configured at 2.36in. They led to equally Homeric 2.28in valves.

during the muscle car era. One such new powerplant came on line in 1968. Referred to by insiders as the "385" engine family, the new motor was based on modern "thin wall" casting technology and was built around a short-skirted block that permitted sufficient room for a truly impressive number of cubic inches. In its first iteration, the new engine displaced 429ci and was topped by a set of cant-valved (semi-hemi) head castings that were light years ahead of the old "FE" 427 technology. The new 429s first saw duty in the luxo-land yacht Thunderbird line and then grew to 460ci for use in the even more pachyderm-like Lincoln line. Even so, providing the necessary vacuum to operate all of a Lincoln's power options was not all that Ford execs had planned for the new engine family. Not by a mile (or, actually, 500 miles).

While base 429s and 460s were hard at work providing pedestrian motorvation for the lumbering Lincoln and T-Bird car lines, Ford engineers were hard at work on a street legal, Hemi-headed version of the engine for use on the NASCAR circuit. By late 1968 work towards that end had progressed sufficiently to the point that a number of running prototypes had been built and evaluated. As originally configured the new (and as yet nameless) Hemi engines featured beefy block castings that incorporated four-bolt mains on four of the engine's five bearing throws. A forged steel crank and massive 1/2in bolt-equipped rods were also part of the racing package as was a block-mounted solid lifter cam. A set of domed forged aluminum pistons and a dry sump oil pan system rounded out the short block's "appointments."

As you might have guessed, the engine's new head castings caused the most commotion. In prototype trim, those heads were poured from cast iron rather than lightweight alloy. They featured a set of eight true hemispherical combustion chambers and huge oval-shaped ports. Valve size was equally "Homeric" with intakes measuring in at 2.37in and exhaust at 1.90in. Dual and single four-barrel intakes were developed for the engine and both were cast in aluminum. Even so, cast iron-headed prototype 429 Hemis tipped the scales (crushed them, actually) at over 900lb. Their hefty weight aside, dyno runs of the new engine promised great things in the horsepower department.

R&D work ultimately resulted in a set of aluminum alloy castings that took the place of their oh-so-heavy iron forerunners. In final form those heads featured a twisted or staggered valve arrangement that led the new engine to sometimes be referred to as the "twisted Hemi." Others within the Ford hierarchy took to calling the new Hemi the "Blue Crescent 429" (due to both its Hemi-shaped combustion chambers and Ford's blue corporate racing livery) while still others referred to the engine as the "shotgun." It was at this juncture that engineers began to search about for both a production-based home for the new engine and one universally applicable name to call it. And that was just about the time in 1968 that Semon "Bunkie" Knudsen signed on as Ford's new boss.

As already discussed in the Boss 302 chapters in this work, Knudsen was an inveterate racer who had been a strong proponent of factory backed motorsports during his many years at General Motors. As a result, he greeted the new Ford engine with open arms and immediately became involved in nurturing it to racing maturity. Knudsen's biggest part in that process was the role he played in deciding where to

Boss 429 engines did NOT use conventional head gaskets. Instead, cooper rings and O rings were used to seal in combustion and fluids. Note the full Hemi combustion chambers in this race spec head.

Street Boss engines used heads that were not fully hemispherical in configuration.

"plant" the new monster motor for homologation purposes. Even in final, alloy-headed form, the new Hemi engine was no lightweight. And it was big, too. It was so massive that it couldn't be bolted into just any Ford or Mercury chassis. Original plans for the engine called for it to be installed in the full-sized Ford line as attested to by the "C9AE" (or 1969 Galaxie) casting numbers that the new engines head and block carried for identification.

The only trouble with that plan was that, by 1968, Ford Galaxies were anything but sporty. Though at one point in Ford racing history, the Galaxie line had been the NASCAR standard bearer, by the end of the sixties, Ford's new big car line had evolved into a block-long blunderbuss that was best suited to police car work or "blue haired" transport. And that was a fact that Knudsen grasped immediately. Another obvious "port" for the new engine to call home might have been the Fairlane and Montego lines that were ultimately destined to venture into Grand National battle under the Hemi's estimable power. The problem with that plan was the narrowness of the Ford intermediate engine box. Unlike cars in the full-sized line, Ford intermediates of the day featured front suspensions with coil springs located above the upper control area. As a result, a fair amount of underhood space had to be given over to shock towers. Towers that were directly in the way of the new Hemi's huge head castings and valve covers. Making the new Hemi work in a Torino or Cyclone was possible, but not without major surgery.

Amazingly, and for reasons still not entirely clear, Knudsen ordered the new racing big-block engine to be homologated in the Mustang and Cougar car lines. The Mustang and Cougar lines were both smaller and lighter than their intermediate brethren. As a direct result, their engine bays were even more cramped than those under the hoods of Torinos or Cyclones. Fitting a Boss 429 in a Mustang chassis was sure to be a major undertaking. Repeating that process the 500 times necessary to homologate the engine for NASCAR competition was a daunting task indeed. Still, that's just what Knudsen wanted—and just what he got!

Ford engineers began to investigate the feasibility of installing a Blue Crescent engine in a Mustang chassis in September of 1969. Engineers in Ford's Special Vehicles Division were assigned to the task and shortly after accepting that responsibility, they "contracted in" the staff of Ford racing subcontractor, Kar Kraft Engineering in Brighton, Michigan. Since regular assembly line production was not suited to the special manufacturing steps necessary to make Knudsen's proposed swap work, it was felt that Kar Kraft's specialty facility was just the place to work on the project. The engineers at Kar Kraft were responsible for much of the work that went into the Ford GT program and had also developed several Ford-based prototypes and pre-engineered race cars. The Kar Kraft engineers' work with Ford's Mustang and Cougar Trans-Am teams made them intimately familiar with those two chassis.

Amazingly, the Kar Kraft team had a prototype of the new Hemi-powered Mustang up and running in just three weeks. Incredibly, they accomplished that goal while simultaneously developing the prototype version of the Boss 302 Trans-Am chassis! By September of 1968 Ford executives decided to go ahead with

Black Jade was one of the five standard colors available for Boss 429s in 1969.

the Hemi-powered Mustang homologation program, the production feasibility of that project having been proven by the work conducted at Kar Kraft. It was sometime during that time frame that Larry Shinoda (a Knudsen protégé who'd accompanied Bunkie from GM to Ford) convinced the powers that be at Ford to call the new version of the Mustang the "Boss." The small-block Trans-Am car became the Boss 302 and the Hemi-headed, big-block engine package that Kar Kraft was working on became the Boss 429.

Ford's decision to build the Boss 429 Mustang was undertaken with two goals in mind. First, to homologate the car for Grand National competition. The second goal was to produce a car capable of equaling or besting the street performance of a 426 Chrysler Hemi-powered car.

The first steps taken by Kar Kraft were to locate and outfit a production site large enough to build about 500 cars per year for two years.

Point man for the project at Kar Kraft was Roy Lunn. Lunn had earned his "spurs" at Ford while

Shoe horning a Boss 429 engine into a Mustang engine box required nothing short of major surgery. New and radically widened shock towers were needed to make everything fit.

working on the original Mustang I program and later had played a major role in the world-beating Ford GT-40 (MkII and MkIV) program. Fran Hernandez was the plant manager for the Boss project and like Lunn he'd been involved in preceding years, with several of Ford's factory backed racing efforts—among then the 1967 and 1968 Ford Trans-Am teams. Ford engineer Chuck Mountain was the man responsible for the engine side of the equation and he also played a role in the development of the racing intermediates that the Boss motor would actually race in once homologated. It was these three talented men who brought the Boss 429 project from the drawing board into actual reality.

Boss 429 production began in December of 1968 when an order was placed by Kar Kraft for 100 Mach 1 Mustangs from the River Rouge assembly line. Those cars, ordered in lots of 20 Raven Black, 20 Royal Maroon, 20 Candy Apple Red, and 20 Black Jade (the four Boss 429 colors) rolled off of the Dearborn line powered by 428 Cobra Jet engines. Ford saved money on subsequent Boss 429 conversions by outfitting them with standard mule engines designed just to power them off the line. Once at Kar Kraft the cars were first relieved of their assembly line power plants then modified as per the instruction in a 35-page change order promulgated by Ford. Extensive work was then performed on each chassis' shock towers. Work so comprehensive, that each of the cars' front suspensions and shock tower assemblies was just about completely removed and re-engineered. The purpose was, of course, to hack out enough room to accommodate the massive 30in width of a street Boss 429 engine. No small task when you consider that a stock 1969 Mustang's engine bay was at most 28in wide.

Once a "Boss to be" Mustang's front underpin-

A significant part of a Boss 429s appeal was the way the car's NASCAR-bred engine filled the engine bay—completely. Conservative valve lift, a relatively small 735cfm carburetor, and a rev limiter all conspired to choke off the engine's heavy breathing potential. With a few modifications, the engine came to life. Stock, 428 Cobra Jet-equipped Mustangs ate Boss 429s for lunch.

nings and shock towers had been set aside or discarded, specially modified replacement shock towers were installed. Some of the pieces needed for that conversion were provided by Ford while others were fabricated in-house at Kar Kraft. Extra bracing accompanied the new and wider modified shock towers into each chassis in an effort to ensure that the cars would retain their structural integrity when burdened with a Boss 429 engine. The resulting relocation of each car's front suspension mounting points moved upper control arms outward a full inch. And that in turn required the lowering of those mounting points (a la Carroll Shelby's early GT350s) and the use of re-engineered control arms to maintain correct steering geometry. Part of the Boss conversion process also included the installation of larger than stock diameter spindles designed to handle the extra weight. Stock diameter 11.3in floating caliper front discs remained unchanged from their earlier Cobra Jet incarnation. One final engine box modification was the installation of a Shelby-perfected export brace that was configured to match the Boss 429's unique shock towers.

Once each new Boss' front underpinnings were in place, the NASCAR-inspired engines they'd been modified to accept could be installed. Even with the extra room created by major metal surgery on the Kar Kraft line, mating the 429 Boss engine with a Boss chassis still required the use of new motor mounts that both raised the engine an inch and moved it forward a similar amount (as compared to stock 428 CJ location). Neither of those modifications (or the re-engineering of the front suspension) improved handling. But, of course, sparkling sporty car handling was not what the Boss 429 was about.

Brutal straight-line acceleration (and, of course, NASCAR homologation) was the Boss 429's primary

Boss 429s were pretty much a total package deal in 1969 and few options were available. Mostly because each one of the cars came fully decked out in just about every available option to begin with. In that number were Magnum 500 rims, deluxe interiors, a front spoiler and, of course, the monster engine under the hood.

Boss 429 interiors were "top cabin" for 1969 and included just about every comfort and convenience option available for Mustangs that year. Air conditioning was made conspicuous by its absence, however.

Choke and Ram Air operation was governed by a pair of knobs that were mounted beneath the driver's side dash panel.

mission in life. Or at least that was seemingly the case. Which makes the decision to significantly detune the Boss engines slated for street duty a bit surprising. But, perhaps we are getting ahead of ourselves.

Other changes performed on each Boss chassis as it rolled down the mini-assembly line in Brighton included the relocation of the battery to the trunk (via the use of a special kit that then became a dealer option for all Mustangs in 1969). Each new Boss was also fitted with a special thin cross section power brake vacuum reservoir designed to clear the 429's massive valve covers. One other special modification was the installation of a core support-mounted, auxiliary oil cooler plumbed to the engine via special high pressure lines and an alloy, AN fitting-equipped oil filter adapter.

Close-ratio Top Loader four speeds were bolted to the bell housings of each Boss 429 engine just prior to their installation in a chassis. Beyond this last addition, most of the rest of a Boss 429's drive line and chassis went untouched on the Kar Kraft line. One notable exception to this was the .62in diameter rear sway bar fitted to each car. Working in concert with the .94in front bar, the rear stabilizer was designed to combat the unavoidable understeer produced by increasing a car's front weight bias. Boss 429s also carried the staggered rear shocks that were part of the standard transmission package found under all "stick" Mustangs in 1969.

Once the massive new engine was bolted home, the Kar Kraft crew installed a specially modified hood fitted with a flapper valve-equipped fiberglass scoop.

Unlike the Ram Air systems found on other Ford high performance engines in 1969 and 1970, a Boss 429's forced air setup was manually controlled by an underdash knob in the cockpit mounted next to the car's manual choke knob. Beyond the installation of that duo of interior control knobs and the replacement of all Mach 1 designation (with a "plain" M-U-S-T-A-N-G labeled dash plate) a new 1969 ½ Boss 429's control cabin went unchanged from its plush Mach 1 Dearborn assembly line configuration. A Big Boss' interior came decked out in simulated teak dash and door inserts and covered with yards of black comfort weave vinyl in 1969. An in-dash tachometer, high-backed bucket seats, a floor mounted console, deluxe seat belts, and the interior decor group were all standard Boss 429 "options" in 1969. The first Boss 429 rolled off of the Brighton assembly line on January 15, 1969. By the time that 1969 Boss production ground to a halt, 857 Boss 429 Mustangs had been built. Each carried special "Z" code (5th digit of the VIN number) engine serialization and a driver's door tag designating the car a NASCAR homologation variant.

As already mentioned, the engines Ford shipped to Kar Kraft in Brighton for installation in street going Boss 429s were most assuredly not delivered in anything approaching race tune. Also of interest is the fact that there was more than one version of the basic Boss

1969 Boss 429 Road Test Results

Car Life magazine, July 1969
  zero to 30.....3.2sec
  40.....4.3sec
  50.....5.8sec
  60.....7.1sec
  70.....8.6sec
  80....10.0sec
  ¼ mile 14.09sec@102.85mph

Road Test magazine, September 1970
  ¼ mile 14.43sec@98.68mph

An ebony Boss 429 gave a whole new meaning to the expression, "Bad in Black." Especially when the car was fitted with an optional rear spoiler. Model identification was fairly low profile for Boss 429 Mustangs. In contrast to the flashy side stripes found on Boss 302s, Big Boss cars carried only these low profile decals.

engine. The first 279 Boss Mustangs built, for example, were "S" code Hemi engines. That is to say, engines that had been assembled with a curious mixture of battleship strength, 1/2in rod bolt-equipped full NASCAR connecting rods and 429 CJ "grind" hydraulic camshafts. NASCAR-style magnesium valve covers were also part of the "S" package.

The next Boss Mustangs to roll off of the line were equipped with "T" code engines. These particular Bosses were equipped with 429 CJ style, 3/8in bolt diameter connecting rods instead of the much heavier (and slower to rev) NASCAR forgings. The first handful of "T" motor equipped cars also came equipped with juice cams and Magnesium valve covers while later engines featured slightly hotter solid lifter cam shafts and aluminum covers. "T" engines were ultimately used in most RPO 1970 Boss 429s as well. But not all, as some 1970 Big Bosses were built with "A" code Hemis. In essence, "T" code engines with slightly revised smog equipment.

In every case, street Boss 429 engines came factory-equipped with high-rise dual-plane single 4V intake manifold and a modest 735cfm Holley carburetor. Exhaust chores were delegated to free-flowing "header" style cast iron exhaust manifolds that mated to a stock 428 CJ style dual exhaust system. Ford sales literature claimed the street Boss 429 engine produced 375hp at 5,200rpm and 450ft/lb, while solid lifter-equipped "T" motors were said to produce an additional 25 ponies.

In sharp contrast to the major changes that went on beneath a Boss 429's bonnet, the car's cosmetics went largely untouched. Boss 429s were fitted with a large and fully functional hood scoop unique to the big-block Boss and a gravel pan chin spoiler. Although quite similar to an aero-aid that was part of the Boss 302 package, a Big Boss car's spoiler was actually made slightly "shorter" to compensate for the car's 1in lower than stock stance. The rear deck-mounted spoiler was optional on Boss 429s as were the racy, sports slats for the back light.

Interestingly, 302-style side stripes were not a part of the standard Boss 429 package. Model designation was provided, instead, by a set of low profile B-O-S-S 4-2-9 fender decals that could be easily overlooked. One final component of the Boss 429 package was a set of flashy 15x7 Magnum 500 rims that featured special "tall" center caps (as compared to the hub covers found on Boss 302 Magnums). On balance, the rest of a Boss 429s silhouette was unchanged from its original Dearborn assembly line form. And that meant that all 1969 cars still retained the simulated side scoops and "C" pillar medallions that ordinary 1969 fastback Mustangs carried (save for Boss 302s, of course). That being said, the 1969 Boss 429 was still a fetching sight. And that was before its hood was opened. Popping the bonnet revealed the biggest and baddest-looking regular production engine to ever grace an engine bay and was guaranteed to wow the troops at any Saturday night burger stand. Unfortunately, a Boss 429s performance didn't always measure up to the visceral image the engine presented. In street legal trim, at least. In fact, some road testers of the day went so far as to call Ford's Big Boss "a stone." The car's road handling was often taken to task, for example. That's not at all surprising when you consider the fact that fully 57% of a Boss 429s 3,670 lb. curb weight rested squarely on its front tires. And, as any sporty car type will tell you, front weight bias works best at producing smoky, tire howling understeer in the twisties. That being said, however, some car scribes did note that the Boss 429 understeered less than, say, an equally beak heavy 428 Cobra Jet Mustang. All the weight on a Boss' nose produced superior braking; a Big Boss could generate more than 1G in deceleration.

Even more disappointing than a Big Boss' somewhat indifferent handling was its lackluster acceleration. *Car Life* tested a black Boss 429 in July of 1969 and found that quarter-mile times below the 14sec mark were not possible in production trim. Though the car's 102mph terminal velocity at the end of the 1320 was promising, the 14.09 that accompanied it was significantly slower than the 13sec performance of

The Boss 429 made its racing debut at the Atlanta 500 in March of 1969. Here Glen Wood (one half of the Wood Brothers) sits on one of the engines (perhaps THE engine) that Cale Yarborough used to score that win. Daytona Racing Archives

a stock 428 CJ-powered pony car. A January 1970 Boss 429 thrash conducted by *Car Craft* produced similarly uninspiring times in the 14.08/106mph range.

So, just what was the problem with Ford's biggest Boss in regular production trim? The answer is a combination of things, all of which were probably attributable to both the hurried pace of the Boss 429 homologation and the very limited nature of its production run. Take for example, the "juice" cam that came stock in the very first 1969 "S" code engines. Though a Boss' alloy pistons came factory fly cut for cams with lifts in the .600in range, "S" motors sported .509in lift cams that were more than a little on the conservative side. The underhood rev limiter helped choke a Big Boss motor by turning out the lights just as the engine's big ported heads started to wake up. And then there was the 735cfm Holley carburetor that Ford engineers saddled the engine with. Though hard to imagine, that particular fuel mixer was nearly a full 50cfm smaller than the four barrel that came as factory equipment on "baby" Boss 302 engines. Big intake ports and huge valves (like those in the Boss 429) need high rpms and high rpms need lots of cfm to work best. And that type of performance just wasn't possible with the Boss stock cam and carb package. The 3.50:1 differential that came stock in the Boss 429 package didn't do all that much to keep the big engine in the proper rev range around town either. And the battleship weight NASCAR style rods that were used in "S" motors made it harder than it should have been for the engine to reach peak power.

Some improvement in performance was realized when Ford shifted to a hotter, .509in lift/300 degrees duration solid lifter cam in the "T" motors that came on line after the first 200 plus Bosses had been built. The lighter weight connecting rods "T" motors carried also helped in the performance department by freeing up rpm production. When backed by an optional 3.91:1 ring and pinion, a "T" motored Boss was a much more formidable mount. Those in search of the truly sparkling performance from a Boss 429 could find it by freeing up the car's breathing with aftermarket parts . When that path was followed impressive performance was in the offing. The same *Car Craft* article that referred to stock Boss 429s as "stones" for example, went on to note the 12.32/113.49 quarter mile passes the same car was possible of returning after a handful of modifications. So, as you can see, though a street Boss was arguably longer on looks than outright performance, there was a true thoroughbred lurking just below its surface waiting to be given its head. And that's just what happened on the Grand National circuit in 1969.

The Boss 429 engine first tasted victory in Atlanta in 1969. The race- spec Spoiler IIs and Talladegas that made that NASCAR win possible were made possible by a special run of street-going droop-nosed Fords and Mercurys like this one.

The Big Boss Hits the NASCAR Circuit

Since the purpose for the Boss 429 engine was racing, Knudsen and the Ford Motor Company scheduled an early competition debut for the new Hemi powerplant. That unveiling was slated for the February 23, 1969 running of the Daytona 500 in Daytona Beach, Florida. Ford added to the excitement of the engine's introduction by simultaneously pulling the wraps off the new aerodynamically designed Torino the Boss was intended to power.

Work on the new Torino had begun around the time Dodge announced a swoopy new version of the Charger called the 500 in October of 1968. The new Charger featured a number of sheet metal changes designed to slip through the air with less resistance. Little did Dodge racers know when they rolled out the new Charger 500 at Charlotte, Ralph Moody was already hard at work on a Ford counter. Holman & Moody, in concert with Ford engineers Bill Holbrook, Luigi Lesovsky, and Bob Swetak, added a new sheet metal proboscis to a Torino's (and later Mercury Cyclone's) normally billboard erect front body work. The end result was a new silhouette that extended forward an extra six inches toward a narrower and lower termination point. A flush-mounted grille and a radically reconfigured bumper (that served as an airfoil) finished off the cars' revised aerodynamic silhouette. The net result was a car that cut through the air like a hot knife through butter. Ford's NASCAR racing chief, Charlie Gray, decided to call the new car the Torino Talladega (after Bill France's new superspeedway being built in Northern Alabama). Ford Head Semon "Bunkie" Knudsen signed off on the design and shortly thereafter prototypes were tested at Daytona. Those shakedown runs indicated that the new Talladega trim was good for a significant increase in top speed, even when pushed around the track by last season's 427 Tunnel Port engine. The extra horsepower offered by a race spec Boss made things get big in the windshield even faster and soon Ford racers were looking forward to the 500 with anxious anticipation.

Unfortunately, even though Ford racers arrived at Daytona in 1969 packing Boss 429 race engines, France and the sanctioning officials remained unconvinced that a sufficient number of street Boss 429 Mustangs would be built. As a result, the engine was disallowed for competition at Daytona and would not be permitted to grace a racing grid until at least 500 Boss 429 Mustangs had rolled off of the Brighton line. France's decision did not derail Ford's hopes for victory at Daytona. At race's end it was LeeRoy Yarbrough's 427 Tunnel Port-powered Torino Talladega that crossed the stripe first. And if that win wasn't dispiriting enough for Mopar racers in NASCAR circles, they knew the day would soon come when the new fleet of Big Ts would pack full-fledged Boss 429 Hemi power for good.

That dreaded day (for the Dodge Boys, that is) arrived one month later at the Atlanta 500. Satisfied that the necessary number of Boss 429 Mustangs had finally been built, NASCAR officials gave Ford racers the green light to go Boss racing. Ford upped the ante by unveiling a new aero version of the Mercury Cyclone called the Spoiler II. Like their Talladega siblings, the new Mercurys featured extended beaks designed to lower their coefficient of drag.

Qualifying speeds were nearly four full miles per hour over the 1968 race and Ford and Mercury drivers did their part to up those speeds. Chargin' Charlie Glotzbach's Charger 500 proved to be the fastest during qualifying, but once the flag fell, Cale Yarborough quickly displayed just how strong his Boss 429-powered Wood Brothers' Mercury was by snatching the lead on lap three. He didn't relinquish that ad-

Ford was more than a little proud of the Boss 429 engine. Here Boss drivers (clockwise from right) LeeRoy Yarbrough, Donnie Allison, David Pearson, Cale Yarborough, and Richard Petty pose with Pearson's Boss 429 powered Holman & Moody Talladega. Craft Collection

The 1970 Boss 429s weren't all that much changed from their 1969 trim. But they did come decked out in a wide variety of dazzling colors. Grabber green was one such hue; non-Ray Ban examination of a Grabber colored car was probably unwise on sunny days.

vantage until lap 51 and then charged back to the front just 10 circuits later. And that was pretty much the story for the rest of the race: Yarborough and his Spoiler II out front. Yarborough's red and white number 21 "Sixty-Minute Cleaners" car led 293 of the 334 laps that made up the event. It was an impressive performance. After the race, Yarborough said, "This new Mercury and Boss 429 engine worked like clockwork." A clock that signaled time was up for the 426 Chrysler Hemi engine.

Ford drivers dominated the rest of the season with Boss 429-powered cars. Victory number two came in the Richmond 500 and that victory was followed by 23 more. In that number were triumphs at just about every superspeedway on the Grand National circuit including convincing wins in The Rebel 400 and the Southern 500 at Darlington, the Firecracker 400 at Daytona, the Motor State 500 and Yankee 600 at Michigan and the Dixie 500 at Atlanta. In fact, just about the only high speed event Boss 429 powered Talladegas and Spoiler IIs didn't

dominate occurred at (ironically) the inaugural Talladega race—the only superspeedway race that factory-backed Ford teams did not compete in. The reason for the Boss 429 absence at Talladega was a dispute between most major team drivers and Bill France about the conditions and safety of the new 2.66mi track. When early high speed runs around Big Bill's new 33 degree banked tri-oval produced blowouts and blistered tires by the score, Richard Petty (who was putting in a one-year stint for Ford driving a Boss 429-powered Petty Blue Talladega) and his Professional Driver's Association decided to boycott the event. In their place, France rounded up a motley field consisting of a handful of independent Grand National drivers and a mixed bag of Grand American Mustangs and Camaros. Richard Brickhouse finished first in a race marked by a great number of mandatory caution laps and tire inspection pit stops (scheduled to avert a tire failure caused disaster) and his victory was one of the few bright spots for Mopar racers during the 1969 season.

Though Boss 429 cockpits were still plush affairs in 1970, they weren't quite as cushy as they had been in 1969. And that's because 1970 cars came fitted with semi-deluxe interiors (with standard rather than padded door panels) and not full boat Mach 1 "fixins'." Jim Smart Photo

When the last Boss 429 motor fell silent at season's end, Talladega and Spoiler II drivers won 30 of the 52 races in the series and David Pearson was the Grand National driving champion for 1969. It was a stunning performance made all the more impressive by the fact that Chrysler engineers had grafted radically pointed beaks and soaring rear wings to the Charger 500s (creating the Dodge Daytona in the process) but could not sort out the new cars in time to derail the Talladega/ Boss 429 juggernaut.

After 1969, Ford and Mercury were looking forward to another year of factory backing and competitive racing with the Boss 429 during the 1970 season. Despite Dodge and Plymouth's new winged Daytonas and Superbirds, Ford drivers had plenty of reasons to be optimistic. The new for 1969 Boss racing engine had proven both powerful and reliable. Better yet, those engines were slated to power a new and potentially even faster aero variant called the Torino King Cobra for 1970. Based on Ford's re-bodied Fairlane/Montego intermediate unitbody, the King Cobra featured a dramatically reconfigured front clip that flowed in a nearly unbroken arc from the "A" pillars to the pavement. The new car looked fast on paper and early prototype tests were well underway with Boss 429 power when Ford Head Semon Knudsen received his pink slip from Henry Ford II. Iacocca's ascendance and subsequent decision to slash the racing budget by 75 percent ended the King Cobra project and curtailed Ford's chance of winning the NASCAR crown in 1970.

Ford racers returned to the circuit for 1970 with Boss 429 Talladegas and Spoiler IIs, but lack of the factory backing made the year-old Fords no match for new winged Mopars. Though Ford and Mercury drivers were able to win 10 races, 1970 was a Mopar year on the circuit.

That disappointing season was not the end of the Boss 429 racing glory. Not by a long shot. When NASCAR officials outlawed special aero-bodied race cars in 1971, Boss 429-powered Cyclones and Torinos again proved to be some of the fastest in the series. Bobby Allison, his brother Donnie, A.J. Foyt, and David Pearson all scored Ford victories that season. Most NASCAR fans will recall the incredible string of superspeedway wins that Pearson and the Wood Brothers scored with a Boss 429-powered 1971 Cyclone between 1971 and 1973. Ultimately, NASCAR found it necessary to outlaw seven liter engines like the Boss 429 in the mid-seventies. Speeds had become too fast, the sanctioning body lamented. And the restrictor plates first introduced in 1970, had not proven to be effective in trimming speeds sufficiently. Therefore, the Boss 429 had to go, they said. And so, Ford's Hemi was finally silenced for good. Even so, it's still interesting to speculate about what might have been. More to the point, to wonder about what might have happened to the Boss 429 engine had it not been outlawed two decades ago. When asked that question recently, NASCAR mechanic par excellence, Robert Yates (former Holman & Moody engine builder in the sixties and current owner of the number 28 Havoline Thunderbird), calculated that had Boss 429 engines been given the benefit of 20 years more of development they'd be producing well in excess of 1,000 horsepower in NASCAR-legal tune today. And that would truly be BOSS!

The 1970 Boss 429

Boss 429 racing victories scored on the NASCAR circuit coupled with the excitement generated by Ford's public relations machine made the first street Boss 429 Mustang built by Kar Kraft quite popular with the buying public. To the point that irate potential Boss buyers often found themselves unable to lay their hands on a Boss car at just about any price. Ford was only obligated to build the homologation cars in order to legalize the Boss engine for motorsports use. With demand on the showroom floor high, Ford was forced to build an additional 350 or so Bosses just to keep dealers happy. While high product demand might normally warm a manufacturer's heart, when it came to building Boss 429s, more was definitely less for Ford. And that's because of the high costs associated with the special modifications necessary to convert a production line Mustang chassis to Boss specifications. Fact of the matter was, Kar Kraft billed Ford $4,444 for the conversion work it performed on each Boss Mustang (and Cougar) it built. Add in the engine and regular production costs and you'll see why Ford was reluctant to build very many examples of a car that sold for about $4,000. Truth be known, Ford lost its shirt on every single Boss 429 built and wouldn't have made out even if a Boss 429 had been stickered at $10,000 1970 dollars (Ferrari territory in those days). As a result, it's not likely that demand for the Boss 429 created many smiles in Dearborn. But it did create a second version of the car for 1970.

The actual production period for Boss 429 Mustangs ran from January of 1969 to December of 1969. As a result, Boss production spanned two separate model years, 1969 and 1970. Since the 1970 model year was slated to only cosmetically freshen the Mustang line, extension of Boss production into the 1970 model year did not pose major tactical or mechanical difficulties for the engineers and workers along the Kar Kraft assembly line. When 1969 production at Brighton ended with a candy apple red 1969 Boss late in July of 1969, work on the 1970 version of the Boss 429 resumed at Kar Kraft (after a brief break for River Rouge retooling) one month later.

Like their more mundane 1970 counterparts, second year Boss 429s received a cosmetic makeover. The quad headlight arrangement of 1969 was replaced with a dual lamp layout that featured simulated brake cooling slots in each redesigned front fender extension. A new tail light panel carrying cast metal light bezels was also part of the redo, and unlike the concave tail light panel used in 1969, the 1970 style was perfectly flat. The simulated quarter panel scoops and "C" pillar medallions that designer Larry Shinoda had railed against the year before were both jettisoned.

The functional fiberglass hood scoop was retained, though it was painted low gloss black regardless of a "host" car's base color. Exterior model designation, as in 1969, was limited to a pair of low profile B-O-S-S 4-2-9 decals mounted at midline on the front fenders. Stamped steel, black centered Magnum 500s and the

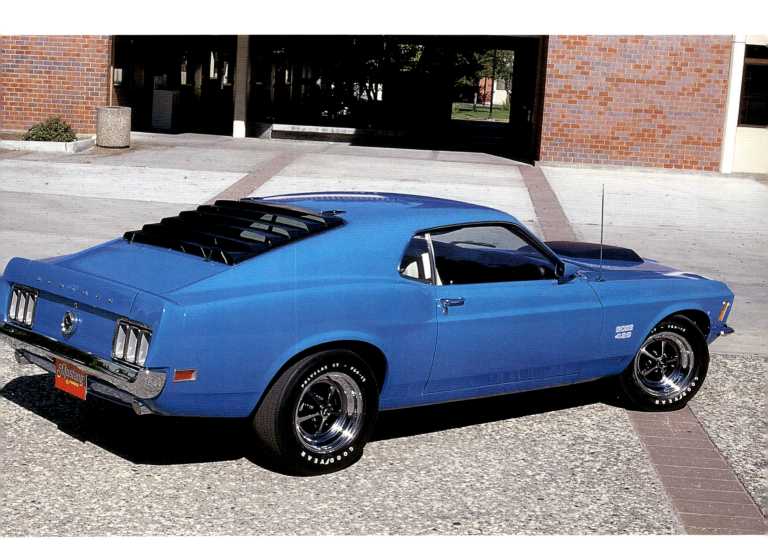

As in 1969, rear deck spoilers were only optional in the Boss 429 line. Many cars came off of the Brighton line without them. Jim Smart photo

91

window shades continued as optional exterior equipment, though not all that many Boss buyers ultimately opted for them, it seems.

What was new for the 1970 Boss 429 line was a veritable rainbow's worth of factory colors. The 1970 Boss cars came dressed up in laser bright colors such as Grabber blue, Grabber green, Grabber orange and Calypso Coral. Buyers with less technicolor tastes could also order up a Big Boss in low-profile pastel blue.

As in 1969, plush was the watch word for Boss cockpits but Ford turned down the sybaritic comfort a bit by only offering the "semi-deluxe" treatment instead of the full Mach 1 interior. The main difference between the two packages was the replacement of the 1969 style, molded door panels (and their wood grain inserts) with a pair of base style vinyl covered panels that carried conventional arm rests. Most everything else in the interior went unchanged between 1969 and 1970. And that included the simulated teakwood dash panels, the comfort weave covered high-back bucket seats, the rim blow steering wheel, and the floor-mounted console. Although down a bit on creature comforts for 1970, a Boss 429 was still a pleasant place to rack up a little seat time.

Beyond the camshaft and rod differences already noted between "S" and "T" versions of the Boss motor, the basic Boss 429 mechanical package remained unchanged for 1970. Aluminum valve covers were used exclusively that year, but things were pretty much the same under the bonnet. Still in residence under the hood, for example, was an auxiliary oil cooler. Ditto for the radically modified shock towers, the unique exsport brace, the rev limiter, the special power brake booster, and the red-coated copper cable that led to the trunk-mounted battery. And, of course, the Big Boss motor itself was also pre-

One thing that wasn't changed for the Boss 429's second year of production was the big iron the cars packed under their hoods. In fact, the "T" motors that 1970 Bosses carried were actually a little hotter than the very first "S" motor 1969s had been. Jim Smart Photo

sent and accounted for in all Boss 429 engine bays for 1970.

The rear sway bar mounted above the axle in 1970 (as opposed to below the axle in 1969), but the suspension componentry was, for all practical purposes, the same.

As you might have guessed, since the basic Boss package went largely unchanged for 1970, so too did the car's performance. And that meant that a 428 Cobra Jet Mustang could eat a Big Boss' for lunch. As in 1969, the severe detuning of the Boss 429 kept it from garnering much respect.

By the time that the last 1970 Boss had rumbled out of the Kar Kraft facility in December of 1969, a grand total of 499 "Z" engine code Boss Mustangs had been completed. Added to the 857 Boss 429s built in 1969, 1970 production brought Boss 429 Mustang numbers to 1,356. Built at a loss solely to satisfy a sanctioning body rule, Boss 429s represented the high water mark of Ford's corporate commitment to racing; it is doubtful that their like will be seen again.

Grabber Blue was another dazzling new color option for 1970. Jim Smart Photo

FIVE

The Boss 351

Ford racing chief Jacque Passino once referred to Ford's Trans-Am Boss 302 program as the "last rose of Summer, because after that everything dies." And while those sentiments might apply to Ford's racing program, they definitely don't apply to Mustang performance on the street. Although Ford's racing endeavors came to a screeching halt, the high performance Mustangs didn't die off overnight. As a result, the roses of Boss Mustang performance continued to bloom well into the 1971 model year.

Though Bunkie Knudsen's time at Ford produced a great many changes in corporate practices in general and the Mustang line in particular, the 1971 Mustangs that were arguably the most influenced by his high performance predilections. The 1969 and 1970 Mustang model lines were almost fully formed by the time that Knudsen took up residence in Dearborn. Although Knudsen and his cohorts (most notably, Larry Shinoda) did exert a great amount of influence on those cars' final styling packages, they didn't really have much impact on their overall configuration. The same cannot be said of the rebodied Mustang line that was unveiled for the 1971 model year. The Mustangs that debuted that year represented the third major restyling of America's original pony car. Like the makeovers that had occurred before, the 1971 Mustang was longer, lower, and wider. Wheelbase for the Mustang line grew an additional inch, for example, while width expanded three. Total length stretched

The 1971 Boss 351 was the last of the Boss Mustangs; it was also one of the finest. Magazine testers reported mid-13sec quarter-mile times and 0.72g skid pad results. Even taking into account the suspicions that Ford discreetly modified the test cars, the performance was impressive.

out an extra two inches, making the new pony cars larger in just about every dimension. The overall effect was a dramatic evolution of the long-in-the-hood/short-in-the-bustle 1969 and 1970 Shelby Mustang line.

While Knudsen was in charge at Ford and the 1971 Mustang line's styling still in flux, it was assumed that Ford and high performance would continue to be synonymous terms. As a result, many mechanical and stylistic choices were made that were ultimately reflected in the line's final configuration. One of these was the development of a high performance version of the 351 Cleveland small-block engine that was introduced during the middle of the 1970 model year.

Even though the Boss 302 and the 351C shared the same cant-valved, large-port head castings, they had little else in common. Their block castings, for example, were radically different. Whereas the Boss 302's foundation was the Tunnel Port short block, which was essentially a beefed-up version of the familiar 289 block, the new 351 Cleveland featured an integral cast iron timing chain cover and, in Boss 351 form, a rugged four-bolt main journal. Although a forged steel crank was not part of the Boss 351 program (as it had been for the Boss 302) specially Brinell-tested cast cranks were. Shot peened and magnafluxed $\frac{3}{8}$in rod bolt equipped connecting rods worked to keep that crank in touch with eight 11.1:1 compression ratio forged "pop-up" pistons. The 351C also used a 324 degree/.491in lift solid lifter camshaft. Those familiar with Boss 302 engines will immediately recognize the head castings used to top off a Boss 351 engine. Fact of the matter was, save for a slightly revised route for cooling fluids employed by 351 engines, the heads were identical. Which is to say that both featured large free flowing ports, generously sized 2.19in intake and 1.73in exhaust valves mounted in poly-angle semi-hemi fashion and small, kidney-shaped, combustion

95

Boss 351 heads were just about identical to the castings that had been used on Boss 302s the year before. They featured poly-angle valves, screw-in guide rocker studs, pushrod guide plates and generously configured ports. This particular engine uses aftermarket roller rockers. Boss 351 heads live on today in the current crop of Cleveland-evolved NASCAR heads.

The Boss 351 engine package looked quite racy with its Ram Air bonnet and aluminum valve covers. Note the factory rev limiter mounted on the passenger side shock tower.

chambers. Screw-in rocker studs and hardened push rod guide plates were also present and accounted for under a Boss 351's stylish cast alloy valve covers (also basically identical to those used on the Boss 302 line). Unlike their smaller Boss counterparts, though, a 351's heads carried D1ZE-B casting numbers that reflected their 1971 date of origin. Induction chores were delegated to a dual-plane high-rise alloy intake and a 750cfm Ford Autolite fuel mixer. A Ram Air system mated to an underhood plenum force-fed cold air to the carburetor through a NACA-scooped hood. Other bits and pieces of the Boss 351 mechanical package included a set of high velocity, cast iron exhaust headers, a dual exhaust system, a high-capacity oiling system, a dual-point distributor, and a flex fan-equipped, extra capacity cooling system. Ford rated the new high performance engine at 330hp and claimed 370lbs-ft of torque.

Like the previous two iterations of the Boss Mustang, a four-speed manual transmission was the only choice available with a Boss 351 engine in 1971. That particular wide-ratio gear reducer was governed by a Hurst shifter and handed off the engine's output to a stock 3.91 ratio ring and pinion that was carried in a 31-spline, Traction Lok, nodular iron, 9in differential.

Ford referred to a Boss 351's underpinning as the competition suspension option. Higher rate front coils and rear leaves, a stiff ⅞in front sway bar and matching ⅝in lean resister, staggered rear shock absorbers, and a variable ratio (20.2–16.4:1) "saginaw" style steering were all mechanical constituents of that package as were a set of 15x7 Magnum 500 (or stamped steel/dog dish hub equipped) rims and aggressive, F60 profile rubber. A mixed pair of 11.3in single piston front discs and 10x2in rear drums rounded out the package.

In road-ready trim a Boss 351 tipped the scales at a portly 3860lbs and most of that mass—58 percent—

1971 Boss 351 Road Test Results

Car and Driver, February 1971
- zero to 30 ....2.0sec
- 40 ....3.0sec
- 50 ....4.4sec
- 60 ....5.8sec
- 70 ....7.4sec
- 80 ....9.2sec
- ¼ mile 14.2sec@100.6mph

Road Test, March 1971
- zero to 30 ....2.0sec
- 45 ....4.1sec
- 60 ....5.9sec
- 75. ....9.6sec
- ¼ mile 13.98sec@104.0mph

Sports Car Graphic, March 1971
- zero to 30 ....2.8sec
- 60 ....6.6sec
- 100 ....14.7sec
- ¼ mile 14.70sec@96.2mph

rested on the front tires. Even so, the Boss acquitted itself well in the twisties. In fact, when put to the test, a Boss 351 returned the same .71G skid pad performances as the Boss 302. The entire combination worked surprisingly well, especially when one considers the extra 300-plus pounds a Boss 351 carried over its Boss 302 brothers.

Styling for the Boss 351 was in the Boss 429 vein. Which is to say, not all that removed from the "regular" Mustang line. Ford stylists graced the Boss' block-long scooped hood with the same "black out" (or argent depending on base color) panel that was part of the Mach 1 package. Hockey stick side stripes flowed down the 351's flanks, as well. A color-keyed lower body side treatment (again, either black or argent) was also part of the package as were color-keyed front hood and fender moldings. Like the Boss 429, model designation was fairly low profile and limited to a trio of tape B-O-S-S 3-5-1 decals that were mounted on the front fenders and rear deck lid. A chin-mounted

When Bud Moore returned to the NASCAR ranks from his stint as a (quite successful) Trans-Am team owner, he began to field what were essentially Boss 351-powered race cars.

Here's a PR shot of a prototype Boss 351. Like most preproduction cars, this styling exercise incorporates several features that never made it into regular production. The honeycomb back panel is in that number, as are the Mag Star rims and body color front bumper. It's doubtful a Grabber yellow car ever came outfitted with a vermilion interior, either. Craft Collection spoiler was another standard feature of the Boss 351 styling treatment and could be optionally complemented with a rear deck-mounted spoiler just like the one (in most ways) that was first penned by Larry Shinoda back in 1968. Interestingly, Shinoda's rear window sports slats were no longer an RPO option for sportroofed Mustangs, perhaps due to the fact that their exaggerated fastbacked roof lines were nearly horizontal to the road surface when a car was sitting on the straight and level.

Color combinations for the Boss 351 ran the gamut from mild to wild in 1971. And, unlike the restricted RPO selections that were orderable for Bosses in preceding years, 351 Boss buyers could conjure up just about any 1971 Mustang color and a few special order ones to boot. In that number were black, white, bright red, maroon metallic, light pewter metallic, bright blue metallic, dark green metallic, Grabber Lime, Grabber Yellow, medium yellow gold, Grabber Green Metallic and Grabber Blue. As mentioned, the base color determined whether the stripes and panels that made up the Boss styling package were laid down in black or argent.

The control cabin of a Boss 351 could be configured in a wide variety of comfort and convenience levels. Simpler tastes could be satisfied with plain black vinyl high-back buckets and an AM radio, while those in search of sybaritic comfort could order up a plush interior dripping in more vinyl calories than just about any four-wheeled transport short of a Rolls Royce. Comfort weave, color-keyed buckets, molded Mach 1-style door panels decked out in simulated wood grain, a fold down rear sports deck, a tilt steering column, an in-dash tachometer and trio gauge combination, a floor-mounted console with auxiliary glove box, a premium sound system and even (in a first for the Mustang line) power windows were all available to those set on wretched excess. While not the Taj Mahal, a fully optioned Boss 351 came pretty darn close.

Road testing the Boss 351 turned out to be a bittersweet experience. *Car Craft* magazine's competition editor, Ro McGonegal, lamented the fact that the 351 he tested in March of 1971 was "probably the last Boss Ford (would) ever build." Making the car's departure all the more disappointing to McGonegal was the fact that the, "general public didn't know the car was

Another shot of the same Boss 351 prototype reveals several other non-production features. Note the non-stock scoopless hood. Craft Collection

Like all Boss 351s, this Grabber yellow car features a grille that carried a set of fake fog lights.

The exaggerated fastback roof line found on all Boss 351s angled at 60 degrees. This white (NOT Wimbledon, by the way) Boss sports an optional rear deck spoiler.

The base Boss 351 interior was outfitted in plain (usually black) vinyl like the one in this car. Note the standard door panels and arm rests that were part of the package.

available." According to McGonegal's article, the Boss was introduced with no advance or advertising in November of 1970 and quietly "slipped to the public." When not lamenting the 351's passing, McGonegal and the *Car Craft* testers had warm things to say about the car itself. Especially the 13.70sec, 104.28mph quarter mile passes they were able to click off in their test car. *Sports Car Graphic* and *Hot Rod* magazines also raved about the Boss 351's straight line ability. The *Hot Rod* testers, for example, coaxed their test Boss to an asphalt shredding 13.6sec, 107mph trip down the 1320 simply by unbolting the car's exhaust system. These impressive times should be taken with a grain of salt, as Ford apparently supplied test cars that were better than stock. According to Phil Hall's *Fearsome Fords*, "for some reason, Ford modified the cars that went to the car buff books for testing." Go figure.

Even so, the Boss 351 was a potent package. And that's just what *Car and Driver* road tester par excellence, Pat Bedard concluded in a February 1971 flog. Bedard reported that the Boss 351 offered "drag strip performance that most super cars with 100ci more displacement will envy and...high lateral cornering forces." But all wasn't bliss with the Boss' suspension in Bedard's opinion. Although capable of crisp performance when driven hard, Bedard felt that the car still understeered too much when off the throttle. And he also took Ford engineers to task for a ride that was far too harsh over uneven pavement. Those criticisms aside, Bedard still concluded that the Boss 351 was a high performance touring car of the type that would

Medium Yellow Gold was one of the less frequently ordered "stock" colors available for 351 Bosses.

probably be owned by hard core enthusiasts. Ultimately, 1,806 buyers stepped forward to purchase 1971 Boss 351s. Those buyers were the last to feel the thrill of taking a new Boss Mustang home from the dealer because, after 1971, the Boss Mustang was no more.

In stark contrast to the Boss cars that had garnered glory and acclaim in NASCAR and SCCA circles in years before, there was no official racing program established for the Boss 351 (either as a complete package or racing engine). Trans-Am racing versions of the 1971 body style made it to the planning stages before Ford's racing division folded, but, they never made it to the track. Bud Moore's plans to campaign a pair of 1971 Mustangs on the Trans-Am circuit progressed to the point of picking up a number of (two or three as Moore recalls today) six-cylinder fastbacks that had been built in Dearborn without interiors or sound deadener. Although Moore actually carried those cars back to his Spartanburg, South Carolina shop, he never converted them to race specifications. Instead, he opted to campaign a team of "leftover" 1970 Boss 302s that season. While other racers like Warren Tope did attempt to score racing points with third generation Mustang Trans-Am cars, their efforts fell far short of the mark set by Moore's Bosses in 1970 (Tope's red, white, and blue 1971 Mustang's best result being a

When the optional floor console was not ordered, a Boss 351's Hurst shifter came surrounded by a mini-console like the one in this car.

third in points performance during the 1972 SCCA season). Though gone, Ford's line of Boss Mustangs is definitely not forgotten. Today Boss 302s, 429s, and 351s are the stars of just about every enthusiast event where they show up. And that's not really surprising, is it? After all, the cars are truly BOSS!

When dark colors like bright blue metallic were ordered for a Boss 351, the line's normal stripe kit and black-out treatment came in argent.

SIX

The 428 Cobra Jet Mustang

Although Ford had been in the big block business for more than a half decade when the Mustang debuted in late 1964, the largest engine available for pony car purchasers was a 289ci small block. And that's pretty much the way things stayed until 1967, when the 390ci engine was offered in the Mustang.

The 390ci engine evolved from the 332ci big block first used in 1958. Ford's first big block, obviously, wasn't all that big in the displacement department, but that changed dramatically. The new engine featured a beefy, long-skirted block and an intake manifold that intruded into the valve cover area. Fairly conventional in configuration (in-line valve train, wedge-shaped combustion chambers, 90 degree cylinder angle, etc.) the new engine had plenty of potential to grow. By the end of 1958, Ford was offering a 352ci version designed with high performance in mind. By 1961, displacement ballooned to 390ci and high performance versions of the engine were cranking out serious horsepower.

The increasing size and power reflected the horsepower war that was being waged in Detroit at the time. Ford was a major combatant in that struggle and, in 1962, upped the ante by punching out the 390 to 406ci and adding a bulletproof, cross bolt main journal bottom end and multiple carburetion. NASCAR and NHRA victories came as a direct result of this big block evolution, but the competition was fierce. GM countered with a series of Chevrolet and Pontiac big inchers of its own and, not to be left behind, Mopar was soon campaigning big-block engines as well.

This 1969 Cobra Jet-powered Mustang is equipped with the Ram Air option. Ram Air cars came factory fitted with shaker hood scoops that bobbed menacingly through the hood in tune with every powerful pulse of the engine.

The FE line expanded to a robust 427ci in mid-1963 and, in peak, dual four-barrel trim, was rated at 425hp and probably put out quite a bit more than that. Ford's 427 FE became a legend both on and off the track, winning races and more than a little respect all across the country. Arguably one of the greatest racing engines ever built, Ford's 427 was instrumental in cinching scores of victories from LeMans to Riverside and back again. But it was expensive to build and maintain, and was peaky and temperamental when put to regular road use.

When Ford's Mustang line appeared in mid-1964, there was no big-block engine on its order sheet, primarily because of dimension. The bottom line was that an FE engine just wouldn't fit into the Mustang's narrow, Falcon-derived engine bay, no matter what its displacement might have been. At first, that didn't really matter all that much. After all, the Mustang was the sole example of the pony car class that its introduction had created. So a big-block engine just wasn't needed to beat competition that didn't exist.

But all of that changed in 1967 when Chevrolet rolled out its own pony car, the Camaro. All of a sudden, two long in the hood/short in the bustle sporty cars were vying for the same market segment. Worse yet (for Ford), the Camaro could be factory equipped with a fire-breathing 396ci big-block engine.

Ford countered by sliding a 390ci FE engine into a widened Mustang engine box that was available in mid-1967, but that wasn't enough to counter the potent new Camaro on or off the track. While a 390-equipped Mustang GT was capable of mid-15sec quarter mile times, a 396-powered Camaro made the same trip down the strip a full second quicker. That fact was not lost on the buying public or the motoring press. Soon automotive scribes like *Hot Rod* magazine's Eric Dahlquist were openly sneering at big-block Mustang's inability to best its new Camaro nemesis. Quite

Dyno Don Nicholson's SS/F car was one of the fastest at Pomona. It should have been since the crew at Holman & Moody/Stroppe were responsible for its preparation. Craft Collection

> ## Cobra Jet Test Results
>
> **1968 ½ 428 Cobra Jet Road Test Results**
> *Hot Rod* magazine, March 1968
>   zero to 30 ....3.0sec
>   40.........................3.4sec
>   50.........................5.0sec
>   60.........................5.9sec
>   ¼ mile 13.56sec@106.64mph
>
> **1969 428 Cobra Jet Road Test Results**
> *Car and Driver,* November 1968
>   zero to 30 ....2.1sec
>   40.........................3.0sec
>   50.........................4.4sec
>   60.........................5.7sec
>   70.........................7.2sec
>   80.........................9.5sec
>   ¼ mile 14.30sec@100.00mph
>
> *Car Life,* March 1969
>   zero to 30 ....2.6sec
>   40.........................3.4sec
>   50.........................4.4sec
>   60.........................5.5sec
>   70.........................6.9sec
>   80.........................8.4sec
>   ¼ mile 13.90sec@103.32mph
>
> *Popular Hot Rodding,* January 1969
>   ¼ mile 13.69sec@103.44mph
>
> **1970 428 Cobra Jet Road Test Results**
>
> *Road Test,* February 1970
>   ¼ mile 14.31sec@100.20mph

understandably, Mustang sales began to slip. Dahlquist even went so far as to initiate a write-in campaign for disgruntled Ford High Performance buyers in November of 1967. When the tear-out form that Dahlquist provided in that issue of *Hot Rod* started flooding into Dearborn in the thousands, Ford execs began to get the message.

Something had to be done, and fast. Carroll Shelby was one of the first in the Ford ranks to step into the horsepower breach with a response to the big-block powered Chevrolet Camaro. Shelby's initial answer was to make a raid of the Ford parts bin in an effort to create a big-block Shelby Mustang based Camaro fighter. Shelby's first stop was the Galaxie/Thunderbird parts bin where he stole off with a supply of newest 428ci iterations of the FE line introduced in 1966 to power luxury cars and police interceptors. Smaller in bore size and longer in stroke than the race-bred 427 engines, the new 428s were more manageable in street trim than the peaky 427. Better yet, they were a lot cheaper to build and service since they didn't rely on forged steel cranks, solid lifter camshafts, cap screw connecting rods, or any of the other exotica that was part of the 427 package.

Shelby topped off the 428s he had shipped to his California facility with a dual four-barrel carburetor and a low-restriction, oval air cleaner. He called his new big-block Mustangs Shelby GT500s and they were a partial answer to the Camaro threat. Trouble was a 1967 GT500 was still more than a half second slower through the timing lights than a 396 Camaro. Shelby ultimately toyed with the idea of building a 427 medium-riser powered line of 1967 GT500s, but ultimately backed off of the plan for the same reasons that Ford didn't offer the engine in a Mustang due to high production and warranty costs.

A northeastern Ford dealer named Bob Tasca stepped in to fill the big-block bill for Ford. By 1967, Bob Tasca and his fleet of Ford drag race cars had a high-profile reputation for high performance. That racing success translated into showroom traffic which earned Tasca another kind of respect from Ford executives in Dearborn. As a result, when Tasca asked for engine parts to experiment with, the requested castings were at his doorstep promptly. Rather than rely on fiery 427s to power a fleet of one-off street going Mustangs to sell at his dealership, Tasca chose to bolt a set of early 427-style head castings (referred to as "Low Riser" castings) onto a 428 block. Since the 427 and 428 had sprung from the same engine family, things like bore spacing and cylinder head bolt location were the same. In fact, it was a simple bolt-on affair that produced big results. As it turned out, all the inherently torquey 428 engine needed to wake it up was a set of free-flowing 427 heads. The combination produced an engine with plenty of low-end torque and respectable top-end power to boot. Impressed with performance reports about Tasca's new engine, Ford engineers gave the project closer attention.

Goaded by the 396 Camaro, the Shelby GT500 Mustangs, and Bob Tasca's 427/428 hybrid engine package, Ford decided to create a high performance version of the 428 FE engine of its own. In time, that new engine package came to be called the Cobra Jet 428. Cobra Jet displacement began with the casting of a new fortified version of the luxo land yacht 428 block that featured extra ribbing in the all-important main web area and beefier main caps. Not nearly as strong as the cross bolted, main cap setup used on the 427 engine, the new 428 blocks were still a significant improvement over base 428 castings. All Cobra Jet en-

Here is a comparison shot of the intake ports of the most popular high-performance heads in the FE engine family. From left to right we have the 427 Low Riser head; the 428 Cobra Jet head; the 427 Medium Riser head; the 427 High Riser head; and the 427 Tunnel Port head. Eagle-eyed readers will note that the Low Riser head and the Cobra Jet head are identical.

Ford's Cobra Jet engine made its competition debut at the 1968 Winternationals in Pomona, California. Ford sent a fleet of specially prepared Cobra Jet-powered cars to the track to secure victory in the Super Stock ranks. Unlike the regular production Cobra Jet cars, the fastback that Dyno Don campaigned at the 1968 Winternationals was built from a non-GT chassis that had come down the assembly line without sound deadener or seam sealer.

In 1969, Ram Air was optional. The above 1969 Cobra Jet engine—sans Ram Air—came dressed in chrome valve covers and an equally shiny air cleaner lid. Note the free flowing cast iron exhaust manifolds the engine also wore. The AN fitting equipped oil filter adapter marks this engine as a Super Cobra Jet version. Note the vacuum diaphragm protruding from the right side of the air cleaner. It incorporated a pop-off valve that would open a flap on the air cleaner when the throttle was mashed.

gines were fitted with cast crankshafts that had been selected for their high nodularity. Each was machined to incorporate a 3.984in stroke that, when coupled with the engine's eight 4.130in cylinders, yielded a total displacement of seven liters. In keeping with the "cheap to build" theme of the new power package, Cobra Jet engines also carried an octet of connecting rods that had been lifted from the Police Interceptor engine package. Each of those "base" (more about the not so base, Super Cobra Jet engine momentarily) Cobra Jet rods was fitted with a pair of $13/32$in rod bolts that still confound mechanics with a limited number of sockets. Cast, slipper-skirted, flat-topped pistons were fitted to each rod and they squeezed incoming combustibles at a ratio of 10.6:1.

Valve timing was placed under the control of a hydraulic cam that had originally been installed in the 390 GT engine the Cobra Jet was intended to supersede. Lift on the Cobra Jet/GT shaft (originally listed under Ford part number C60Z-6250-B) was modest at .490in on the exhaust side and .481in for induction. The dual-pattern cam also featured 290 degree duration for exhaust events and 270 degrees for intake chores. Overlap was a modest 46 degrees of crank rotation. By no means radical, the Cobra Jet cam produced a broad, flexible power band and, perhaps more importantly, a decidedly muscular rumble that fairly shouted performance.

As with just about any internal combustion engine, it was the new engine's head castings that determined just how well the overall Cobra Jet package would perform. With a nod towards Tasca's original experiment, Ford engineers topped the new engine with a set of 427 heads. In Cobra Jet trim the heads featured 2.09in intake valves and 1.66in exhaust controllers. They also carried substantially larger intake and exhaust ports than the 390 castings they were intended to replace. Identified by the casting number C80E-6090-N, Cobra Jet heads had conventional, shaft-mounted, non-adjustable rockers and rough cast, 72.8–75.8cc combustion chambers. The principal difference between the revised Cobra Jet heads and the earlier Low Riser castings they were derived from was found in the way their exhaust faces were machined for exhaust bolts (each port carrying four mounting holes instead of just two) and the provisions each head carried for smog pump fittings (in each exhaust port).

Cobra Jet induction chores were handled by a free-flowing but incredibly heavy (90lb) cast iron, high rise style manifold. Flanged for a conventional carburetor (rather than spread bores), the new Cobra Jet casting was identified by the casting number C80E-9425-G. Although putting extra weight exactly where it wasn't needed (over the front wheels of any car it was mounted in), the Cobra Jet intake was still a significant improvement over the low rise, small-runnered 390 GT manifold. In street ready trim, the new intake was fitted with a Holley 735cfm fuel mixer and a single snorkel air cleaner.

Exhaust gases were channeled by a set of low-restriction, header-style, cast iron manifolds that, again, were a significant improvement over the log style manifolds used on the 390 GT engine. Due to their unique bolt pattern, the new manifolds could not be used on other engines in the FE family. An "H" pipe equipped 2 $1/4$in dual exhaust system flowed from the new manifolds to the rear of the car.

The rest of the first Cobra Jet package consisted of chrome-plated, "Powered by Ford" valve covers (replaced in some iterations by ribbed alloy castings late in the 1969 production year), a windage tray that mounted between the oil pan and crankshaft, a vacuum advanced equipped distributor, and a dual groove, alternator/fan belt system. One other significant feature of the Cobra Jet setup for 1968 was the installation of a Ram Air induction system that featured a vacuum diaphragm equipped "flapper" valve air cleaner that was sealed to the underside of the hood with a rubber sleeve. A low-profile fiberglass scoop (a first for the Mustang line) bolted over cut outs in the hood and captured cool air when the car was in motion and admitted it directly to the carburetor under low vacuum (wide open throttle) conditions. The balance of the new engine's drive train consisted of either a big input Top Loader four speed or a specially beefed-up version of the C-6 automatic transmission (that carried a larger second gear servo, a cast iron tail shaft and a host of internal upgrades). Final power transfer to the pavement took place through a nodular 9in rear end and 31-spline axles.

Ford officials, increasingly aware of the chilling effect that high insurance premiums were having on the sales of high performance cars, intentionally low balled the new engine's horsepower ratings in hopes that it wouldn't be saddled with outrageous rates. Despite the new Cobra Jet package's horsepower-enhancing goodies and 38 additional cubic inches, it was rated for exactly the same number of ponies (335) as the 390 GT engine. Truth be known, a Cobra Jet was actually good for close to 400 ponies in factory production trim, and a lot more with minor modifications.

That's exactly what the Super Stock competition found out to their great dismay in February of 1968 at the NHRA Winternationals in Pomona, California. The Cobra Jet's racing debut was made possible by a special run of fastback Mustangs built in advance of regular Cobra Jet Mustang production. The cars, destined to become official team competitors for the Winternationals, were Wimbledon White 2+2s that rolled off of the assembly line sans sound deadener and seam sealer. The cars were then shipped to Ford's Holman & Moody/Stroppe West Coast shop for race preparation. According to a glowing article about those cars published in the April 1968 issue of *Car Craft* magazine, those race ready Cobra Jets single-handedly lifted Ford out of the drag strip doldrums and to a place of prominence.

The SS/E car *Car Craft* managing editor John Raffa tested (with the help of Ford team driver, Dyno Don Nicholson) for that article differed from its street going siblings in a variety of ways. The Cobra Jet under the hood of that race-spec fastback was fitted with a higher lift, long-duration, solid lifter cam. Larger than stock "tubular" sodium-filled, lightweight valves had been

Though most Cobra Jet cars ordered in 1968 were fastbacks, the new big block was also available in both coupes and convertibles.

Four-speed Cobra Jet cars came factory stock with 8,000rpm in-dash tachometers for 1968. Craft Collection

installed, as well. The 1.76:1 ratio, adjustable rocker arms were also part of the SS/E program as was a higher than stock 11.6:1 compression ratio. Taking a peek below the rocker panels was all it took for Raffa to notice the large diameter, tubular steel headers and the sectioned and deepened oil pan installed by the Holman & Moody/Stroppe shop. What wasn't quite so obvious to the naked eye (and definitely NOT known by the sanctioning body at the time), was the significant internal intake and exhaust passage modifications that had been performed to the stock Cobra Jet head castings. Referred to by Ford racing insiders as "Canadian" heads, the Cobra Jet castings that were used on Ford's drag race Cobra Jet engines were actually revised versions of RPO castings that featured head passages and valve bowl configurations that had been greatly revised to increase flow. Ford was able to pull the wool over the tech inspector's eyes by casting up the race only heads with the same ID numbers that stock Cobra Jet units carried and then delivering them to select teams clandestinely (in hotel parking lots in

Ford stylists created a show car they called the Mach 1 in 1968. The car created quite a stir on the show car circuit. It wasn't the last time that the car buying public got excited about a Mach 1 Mustang. Craft Collection

the dead of night!).

Other visible modifications the Nicholson's car noted by the *Car Craft* article included a pair of tubular traction bars that had been welded to the nodular, 9in rear end and mated to the chassis through a set of brackets welded to the frame under the passenger area of the car. Nicholson's car also featured a trunk-mounted battery and a pair of sticky 10.50x15 Goodyear racing slicks.

Nicholson and teammate Hubert (Georgia Shaker) Platt turned in a string of 11.62sec, 119.7mph runs down the 1320 that impressed Raffa, who concluded that, "Ford's new 428 Cobra Jet Mustang Super/Stocker can best be described as a car with hair!" It's more than likely that the Brand X competition found Ford's seven-car Cobra Jet team more than a little hairy, too! Leaving nothing to chance, Ford secured the services of Nicholson, Platt, Gasper (Gas) Rhonda, Ed Terry, and Al (Batman) Joniec to pilot the fleet of Super Stock Cobra Jets in the Winternationals; each of those drivers was a bona fide star in the NHRA ranks at the time. Their driving skill backed by the Cobra Jet's impressive power output put a 428 Mustang in just about every final round face-off in the Super Stock division. Joniec ultimately took top honors in the Super Stock category with a 12.12sec, 109.48mph trip down the 1320. When the bleach box smoke lifted at Pomona, Ford's new Cobra Jet Mustang was parked squarely in the middle of victory lane. That performance was not lost on either the motoring public or the automotive press.

Former Ford nay-sayer Eric Dahlquist became one of the Cobra Jet's earliest and most vocal converts soon after he tested a Wimbledon White, 428 fastback in the March 1968 issue of *Hot Rod* magazine. Like the racing Cobra Jets that had laid waste to the Super Stock competition the month before, Dahlquist's test car had rolled off the assembly line in "lightweight" trim and had then gone to Holman & Moody/Stroppe for a little pick-me-up. Even though dressed out with smog equipment and a full interior, the car was fast. How fast, you ask? Try 13.90sec,

The 428 Cobra Jet Mach 1 was the top of the line expression of performance and luxury for 1969 and 1970. Ford PR types loved to picture the new Mustang Mach 1 in race track settings. Though not actually a race car, a 428 Cobra Jet Mach 1 was the last word in grand touring. Craft Collection

103.96mph run the first time out of the box. Dahlquist's 3.89-geared car backed that run up a few minutes later with an even faster 13.56sec, 106.64mph pass that had him searching for superlatives. "The mere fact that these Dearborn rocket sleds are coming off the production line deserves some kind of award." He went on to say that, "the strength of a single Cobra Jet blast off will put thousands in orbit for the nearest auto loan department." Heady praise indeed—especially since it came from one who had been so critical of Ford's muscle car efforts just a few months before.

Dahlquist's predictions were correct. The Cobra Jet's favorable publicity and fast quarter-mile times had buyers clamoring for Cobra Jet-powered Mustangs. Ford announced the engine's availability in a March 29th announcement to all it's dealers. As outlined in that release, the new engine would be available in all Mustang models (fastback, coupe and convertible). When ordered, the engine would be equipped with Ram Air induction, unique competition handling suspension (which included staggered rear shock absorbers for four speed cars), F70 wide oval tires, a 3.50 axle ratio (with 3.91 and 4.30 gears being optional along with a Traction Lok differential), and a wide black center paint stripe extending from the front to the rear of the hood. The new engine was not available with air conditioning or the optional louvered hood.

When the Cobra Jet option was selected, a 1968 ½ buyer was also required to opt for the GT equipment group (which included styled steel wheels, fog lamps, side body "C" stripes, GT fender and gas cap badges, and "dual quad" exhaust tips) and power disc brakes. In four-speed applications, an in-dash 8,000rpm tachometer was also an option. Suggested retail for the basic engine package was $420.96—a fairly sizable figure in 1968 dollars but still one whale of a performance deal on balance. All told, upwards of 2,800 buyers checked off the Cobra Jet order box during the last half of the 1968 model year. According to the best records available today, 2,253 fastbacks, 564 coupes and 10 or so convertibles were built with Cobra Jet motorvation that year. All of those cars were serialized with an "R" engine code as

The Cobra Jet engine option was available in plush and not so plush forms. This dog dish hub cap-equipped coupe was built with just one apparent option: a Ram Air, Super Cobra Jet engine. But what an option it was!

110

the fifth digit in their VIN numbers. In that number were two different runs of lightweight (sound deadener and seam sealerless) Cobra Jet fast backs and a handful of California Special and High Country Special coupes. Based on the cars from that production run that survive today, it would appear that most 1968 ½ CJs were ordered in either Wimbledon White or Royal Maroon paint schemes and few deviated from the bare bones, base interior trim. Both the nature and number of Cobra Jet production would change dramatically in 1969.

The 1969 Cobra Jet Mustang

So heady had been the praise heaped upon the Cobra Jet engine package in 1968 that there was little doubt about its reappearance in 1969. In keeping with Ford's custom of redesigning car lines every two years, 1969 marked the introduction of a completely restyled Mustang line. Larger in every dimension and sporting a variety of fetching sheet metal changes, the new 1969 Mustang was a high performance grand touring machine in the finest European tradition; especially when outfitted with a Cobra Jet motor and smothered in luxurious Mach 1 ergonomics.

The basic 428 Cobra Jet mechanical package was unchanged for the new model year. As in 1968, a standard Cobra Jet big block came equipped with a hydraulic 390 GT cam, a brace of Police Interceptor rods, cast 10.6:1 pistons, and a single four barrel-equipped cast iron intake. Chrome-plated, stamped steel valve covers and a chrome-plated "spoke pattern" air cleaner lid brightened up the underhood scenery. A chrome-plated dip stick handle completed the effect. Unlike 1968, Ram Air induction was not a

Although Shinoda-style wings and spoilers were not standard equipment in the Mach 1 package, they were definitely available. So were the sport slats that Shinoda also cooked up.

standard part of the Cobra Jet package. Base Cobra Jet engines came equipped with a single snorkel air cleaner that carried a vacuum diaphragm governed "pop off valve" designed to admit more air under full-throttle. Interestingly, Ford's official rating for the engine remained constant at 335hp, even with the Ram Air featured deleted.

Base versions of the Cobra Jet engine were available in all Mustang body styles for 1969 (sportsroof, coupe, and convertible) and could be dressed up in a wide variety of trim levels from dog dish hub cap/bare bones to wretched Mach 1 or Grande excess. In every case, non-Ram Air Cobra Jet cars were serialized with a Q as the fifth digit of their VIN numbers.

Performance-oriented types had the option of ordering up Ram Air induction for their 428 powered Mustang for 1969. And those in search of face-distorting acceleration could also select what has come to be called the "Drag Pack" Super Cobra Jet optional axle ratio package. Either way, Cobra Jets so equipped were identified by the "R" they carried as the fifth digit of their serial numbers.

The Ram Air package for 1969 (and 1970) consisted of a special air cleaner assembly that incorporated a cast alloy air scoop which actually protruded through a hole in the hood when the bonnet was closed. Essentially an evolution of the forced air system first introduced the year before, the new setup featured a vacuum diaphragm operated flapper valve that was situated directly beneath the cast scoop. Without a doubt, the neatest (or should we say the Bossest) aspect of the new induction system was the way it bobbed and shook in response to throttle inputs and idle variations. It was perhaps only natural that the new system came to be called the Shaker. As noted, though the salutary effect of cold air on horsepower production was well known, Ford officials still only claimed 335 horsepower for "R" code CJs in 1969.

The same horsepower figure was listed for Drag Pack-equipped Cobra Jets the same year. Never actually referred to as such on a Ford order sheet, the "Drag Pack" Super Cobra Jet option was conjured up whenever a performance-oriented buyer selected either the 3.91:1 or 4.30:1 ring and pinion and had it mounted on a Traction Lok or Detroit Locker center section. When that occurred, good things started to happen on the assembly line. First and foremost was

The last year for the fabled GT package was 1969. Only a handful of GTs were actually built and the option was dropped a few months into the production year. The GT package consisted of rocker panel striped, styled steel rims, and a GT-logoed, pop-open gas cap for 1969.

the installation of the same super heavy duty, cap screw connecting rods that Ford had developed for use in the Ford GT MkIV LeMans 427 engine. Referred to as LeMans rods, in Cobra Jet trim those beefy forgings featured $7/16in$ rod bolts with heads that had been trimmed for reciprocating clearance. Next came sturdy forged aluminum pistons that, like the LeMans rods they worked in concert with, were designed with the rigors of high rpm work in mind. The extra weight of this new piston and rod combination required an engine rebalance and the installation of an external counterbalancer just behind the front vibration damper.

The rest of the regular Cobra Jet engine package—including the 390 GT derived juice cam—remained unchanged in Drag Pack Super Cobra Jet trim. But the beefed up engines did come graced with racy, finned alloy valve covers and more often than not also sported "shaker" Ram Air systems. One other underhood change that was specific to "optional axle ratio" was the addition of a core support-mounted auxiliary oil cooler that was plumbed to the engine via a pair of high pressure lines and a special AN fitting equipped oil filter adapter.

As in 1968, Cobra Jet and Super Cobra Jet engines could be backed up with either a bulletproof Top Loader four-speed manual transmission or a heavy-duty version of the C-6 three-speed automatic. A race proven "N" case differential and wrist thick 31-spline axles were also part of the Cobra Jet/Super Cobra Jet mechanical menu. So, too, were staggered rear shock absorbers in four-speed applications. As in 1968, an "H" pipe-equipped dual muffler system took care of spent hydrocarbons and produced a delicious basso profundo rumble. One final modification specific to the 428 engine package was the installation of "Boss" style shock tower reinforcement plates that were designed to keep the car's upper "A" frame mounting bolts from pulling through under hard cornering.

As mentioned, a 428 engine could be ordered up to power a wide variety of Mustang body and styling packages in 1969 and 1970. High-performance purists in search of the lowest possible factory stock ET often opted for little more than the Cobra Jet or Super Cobra Jet engine package, and the end result was a plain Jane pony car that looked more like your Granddad's grocery getter than a high performance car. Until the

Race tracks in Michigan, Texas, and Georgia relied on 428 Cobra Jet-powered Mustangs to set the pace in 1970. Craft Collection

throttle was massaged with a heavy right foot, that is.

The other extreme was the super plush Mach 1 styling and interior package. The Mach 1 was intended to serve as a Thunderbird surrogate of sorts and consisted of a number of comfort and convenience options that made the car a hedonistic delight. High-back bucket seats slavered in comfort weave vinyl, plush molded door panels, simulated teak wood dash, door and radio accents, a floor-mounted console, color-keyed rug panels, and a three spoke-rim blow steering wheel were all part of the Mach 1 package.

Other interior options such as air conditioning (available with the base Cobra Jet but not with Super Cobra Jet engines), power steering, power brakes, a tilt-away steering column, a premium sound system, and a folding rear "sport deck" could also be called up to augment the Mach 1's interior treatment.

The external aspects of the Mach 1 package included reflectorized side tape stripes that carried model designations, a low-gloss black hood and cowl, a non-functional hood scoop (or a working shaker in Ram Air applications), a rear deck lid stripe with model designation, a pop-open racing-style gas cap, racing type hood pins and lanyards, and flashy styled steel wheels. Buyers could further alter their Mach 1's cosmetic appearance via the addition of the front and rear spoiler combination cooked up by Larry Shinoda along with rear window louvers inspired by the same fellow.

Car Life magazine called the Cobra Jet Mach 1 package, "the best Mustang yet and the quickest ever" in a March 1969 road test after they were able to coax an "R" code Mach 1 through the quarter-mile lights in just 13.90sec. The Car Life crew found that their test car went "like a hammer in a straight line" and though cursed with a nose heavy weight bias due to the big-block engine under the hood, those road testers still raved that "by choosing the optimum combination of suspension geometry, shock absorber valving and spring rates, Ford engineers have exempted the Mach 1 from the laws of momentum and inertia up to unspeakable speeds." Perhaps most important to high performance buyers of the day was the Car Life statement that "as tested, with two men, test gear and a full tank, all belts tight, air cleaner attached, tire pressure normal, the Mach 1 is the quickest four-place production car ever tested by Car Life." The magazine later named the Cobra Jet Mach 1 its "Pony Car of the Year." Cobra Jet convert, Eric Dahlquist, had moved his typewriter to the Motor Trend offices in 1969 but his enthusiasm for the Cobra Jet engine was undiminished. In fact, he flat out told prospective Mach 1 buyers to skip the base 351 engine option in favor of the "real thing," the 428 Cobra Jet. Popular Hot Rodding added to the Cobra Jet Mach 1's reputation in a January 1969 test by coaxing a 13.69sec, 103.44mph ET from their C-6 equipped NON-Drag Pack test car. The Popular Hot Rodding guys also had good things to say about their test car's handling. In fact, about the only complaint they had to offer was the 30 minutes it took to fuel the car—because people kept coming up to ask questions about it. In sum, the Popular Hot Rodding road test concluded the Cobra Jet Mach 1 package created one of the "sharpest cars on the road today."

As Ford had hoped, the high praise heaped on the Cobra Jet engine by motoring scribes had a direct impact on show room floor sales. By the end of the 1969 model year, 13,193 Cobra Jet-powered Mustangs had been sold. Many of those cars were ordered up with Mach 1 trimming. Since 1969 was the last year for the GT, the handful of 1969 Cobra Jet Mustang GT convertibles and coupes are some of the most collectible Mustangs ever built.

The 1970 428 Cobra Jet Mustang

The final year for FE production in the Ford passenger car lines was 1970. The 1970 Mustangs were essentially identical to the Cobra Jet motorvated cars that had been built in 1969 due to the fact that both cars employed the same unitbody. Styling changes for 1970 included the use of a dual headlamp system in place of the quad lights of the preceding year. Things in the lighting department were also different at the stern where a cast metal affair replaced the three individual (per side) plastic lenses that had been used to signal stops in 1969. Other sheet metal changes for 1970 include the deletion of the superfluous side scoops and "C" pillar medallions that had sullied the flanks of every 1969 sportsroof Mustang. Side marker lamp locations were also revised for all 1970 Mustangs.

Save for those deviations in style, the 1970 Mustang line was pretty much a mechanical duplicate of the cars built one year before. The Cobra Jet and Super Cobra Jet engine was an option for Mustangs of just about every stripe. Fastbacks, convertibles and coupes could all be conjured up with a fire-breathing FE under their bonnets and a wide variety of trim and comfort levels were available. Top-of-the-line ponies were once again called Mach 1s and they continued to represent the last word in high performance Grand Touring. The Mach 1 trim package for 1970 consisted of ribbed lower body cladding that ran the length of the car at the rocker panels. Cast from sheet metal and painted to resemble magnesium, the new panels carried model designation at the base of each front fender. A honeycomb-covered tail light panel and a wide black-out tail stripe distinguished the cars from the rear.

Other cosmetic upgrades for 1970 included the substitution of a narrower black-out hood panel than had been employed on 1969 Machs and in 428 Shaker applications the engine size was spelled out in a thin black stripe that ran the perimeter of the new panel. Fake fog lights took up residence in a new Mach 1 grille and simulated mag style hub caps replaced the styled steel rims that had been standard Mach 1 equipment in 1969. Front and rear spoilers and rear sports slats continued to be orderable on Machs and other sportsroofs.

Interior accoutrements were even less changed than the 1970 Mach 1's only superficially altered exterior. Comfort weave covered high-back buckets were again the central focus and they were surrounded by a rim blow steering wheel, padded teak accented door panels, a floor-mounted console, and a pair of wood grained dash panels. An in-dash tach was optional, too. Base Cobra Jet Machs were again available with AC and power steering and front disc brakes were advisable whenever the big block order box was checked off. The tilt wheel option for 1970 lost its "fly away" function and instead was only adjustable through a vertical five position arc. Ford's old faithful fold down rear seat was also still in residence on the order sheet. As mentioned, a 1970 Mach 1's mechanical and suspension package was virtually identical to the componentry employed in 1969. One notable exception to all that sameness was the addition of a Boss-inspired rear sway bar to help counter the Cobra Jet's understeering proclivities. Finally, 428 Cobra Jet engines also picked up mechanical nannies for 1970 in the form of fender apron-mounted rev limiters.

Non-Mach 1 Mustangs got upgraded to high-back buckets for 1970 and those seats were orderable with plain vinyl, comfort weave, or flashy cloth plaid inserts. A semi-deluxe trim level was also added to the mix and deviated only from the standard Mach 1 treatment in its lack of padded, teak-accented door panels.

Mustang sales were down across the board for 1970 and that decrease was reflected in 428 Cobra Jet production, as well. By year's end, for example, just 2,671 Cobra Jet-powered pony cars had been built. As in 1969, most of those engines had taken up residence in fastback iterations of the line, but a handful of coupes and (a smaller number still of) convertibles were also built with Cobra Jet power. When the last of those rolled off of the assembly line, the high performance FE engine era had come to an end.

This Calypso Coral Mach 1 came factory fitted with a loud vermilion interior. Magnum 500 rims were optional for Cobra Jet Mustangs in 1970.

115

Shelby GT500KR and GT500

Carroll Shelby was the first to squeeze a seven-liter into the engine bay of a 1967/1968 Mustang. That attempt came in 1967 when he built a number of 428 Police Interceptor-powered GT500s. Shelby followed up those first GT500s with a Shelby-ized version of the 428 Cobra Jet he called the GT500KR in 1968 1/2. Most folks agreed it was truly the King of the Road.

In 1968, Shelby Mustang production shifted from the West Coast to the Midwest. Ford had been assuming more and more responsibility for Shelby Mustang production almost since the line had been introduced in 1965. When the lease that Shelby American held on its 6501 West Imperial Highway hangar facility ran out in 1967, the decision was made to shift production nearly all the way across the continent to the A.O. Smith company in Ionia, Michigan. Though that move signaled the beginning of the end of Carroll Shelby's active work on the cars that carried his name, it was far from being the finish of the Shelby Mustang.

When 1968 production commenced at A.O. Smith, new and restyled versions of the GT350 and GT500 models began to roll off of the company's specialty assembly line. As in 1967, big block versions of the Shelby line were powered by slightly tarted-up versions of the 428 Police Interceptor. Just as during the preceding year, those engines weren't really up to the task of powering the increasingly plush Shelby, at least when it came to stop light encounters with 396 Camaros. The GT500's reputation was further sullied early in the model year when a temporary Police Interceptor engine shortage resulted in a number of cars being built with anemic 390 GT engines under their scooped fiberglass bonnets.

Help for the GT500's flagging high performance fortunes finally arrived in April of 1968 when the new 428 Cobra Jet engine became available. Fortunately for Shelby fans, the engine would be available for the 1968 GT500s being built in the last half of the year. Though high production costs prohibited a mid-year styling change to signal the new engine's arrival, Shelby felt strongly that something had to be done to set the car off from the GTs that had gone down the assembly line before. As luck would have it, that was about the same time that Shelby learned that GM was considering releasing a hotted-up version of the Chevrolet Camaro. According to Shelby's GM mole, that new Hi-Po Camaro would be called the "King of the Road," after the popular song by Roger Miller.

Armed with that advance notice, Shelby stole the General's thunder and used the King of the Road name for the new 428-powered GT500. The Cobra Jet Shelby was dubbed the GT500KR. Like the regular Cobra Jet-powered Mustangs that became available at about the same time, Shelby's new GT500KRs

Though more fastback KRs were built than convertible versions, there's no doubt that drop top GT500s were the sexiest of the lot. Simulated mag style hubcaps were standard, but many 1968 Shelby buyers opted for flashier ten-spoke alloy rims.

were well received. An October 1968 GT500KR road test conducted by *Car Life* magazine concluded that Shelby's new Mustang was, "so impressive, so intimidating to challengers, that there are no challengers. The KR breeds confidence bordering on arrogance. The KR driver coasts along, mighty engine rumbling, and looks with a condescending smile on the driver of a lesser car who creeps away from the lights then chuffs past in traffic, knuckles white on the steering wheel." As you might have guessed from that bit of hyperbole, the *Car Life* crew was impressed with the KR's performance both on and off the track and concluded that a 428 Cobra Jet equipped Shelby was well worth stealing—just as their original test car had been! A *Hot Rod* magazine November test flog of both fastback and convertible versions of the KR echoed those same sentiments one month later. As with the *Car Life* test, *Hot Rod*'s Steve Kelly was able to click off mid-14sec ETs with ease in either one of the cars tested. The four-speed equipped fastback GT500KR ultimately proved to be fastest at 14.01sec, 102.73mph and editor Kelly concluded that even more performance might be available with a little carburetor jetting.

According to the Shelby American World Registry, 933 Cobra Jet powered fastback GT500KRs were built in 1968 and they were accompanied down the A.O. Smith line by 381 GT500KR convertibles for a total Cobra Jet-powered Shelby production of 1,314. They have become some of the most highly prized of all Shelby Mustangs.

The 1969 GT500

The final full year for Shelby Mustang production was 1969. Though Carroll Shelby publicly maintained support for the ever encroaching government safety regulations, the savvy old racer could clearly see the writing on the wall. Where once his modified Mustangs had been brash boy racers that made few concessions to the norms of western civilization, by the end of the sixties, the safety freaks were clearly at the gate. Increasing resistance in government and insurance circles to any car in the high performance category coupled with internal competition for the same market segment from cars like the Boss 302 and 429 convinced Shelby that the time had come to pull the plug on his line of modified Mustangs. And so he set up a meeting

In Shelby trim, the Ram Air package dispensed with the need for flapper valves and vacuum diaphragms and just dumped cold air to the engine at all times.

with Ford vice president John Naughton and asked that the Shelby Mustang program be terminated. As history sadly records, the request was granted. But not before the production of what many feel was the best looking Shelby Mustang ever built.

Married to the same two year styling cycle the regular Mustang line followed, Shelby's GT350s and 500s also received a cosmetic update for 1969. And what a makeover it was. Though based on the same basic unit body as the Mustang, Ford and Shelby styling gurus were successful in achieving a vastly different looking silhouette via the addition of an entirely new front clip, a quartet of side and "C" pillar mounted scoops, and a new combination of rear fender extensions and fiberglass deck lid. The overall effect was breathtaking. At the bow, 1969 Shelbys carried dramatically lengthened fenders and a hood that was pierced for no fewer than five different NACA-inspired air scoops. The new grille opening created by this body work was rimmed by a combination of chrome trim pieces and a ribbon-thin bumper, that, while looking good, offered about as much protection as a toothless Doberman. At the rear, new fender extensions and an all fiberglass deck lid created an exaggerated duck tail spoiler. A blacked-out fiberglas rear panel was also installed and it carried a pair of Thunderbird tail lights and a higher mounting point for the license tag. A cast aluminum exhaust port was installed in the original license plate cove and was connected by special ducting to the car's dual exhaust pipes.

Reflectorized body side racing stripes ran the length of the car and carried model designation and a pair of GT-40 MkII-inspired racing scoops. Other bits and pieces of the Shelby trim package included a duo of Lucas driving lamps mounted below the front chrome strip and a quartet of Mag Star five-spoke alloy and steel rims.

With the exception of the radical styling changes, the rest of a 1969 Mustang's mechanical and control cabin resume was lifted directly from the Mustang line with little rewriting. Control cabin changes included a padded roll bar and inertial reel shoulder harness arrangement and a special console insert that carried a pair of analog engine gauges, an ashtray, and a couple of aircraft looking switches. Everything else was pretty much Mach 1 plush except for the Shelby model badges.

Things were pretty much the same under the body work. None of the 1969 GT500's bore the "KR" designation, but they all were equipped with the 428 Cobra Jet engine. Even though the fiberglass of the new Shelby GT package boosted weight to nearly two tons (3850lb to be precise), a February 1969 GT500 road test conducted by *Sports Car Graphic* still produced 14.0sec ETs with speeds in excess of 102mph, which impressed the *Sports Car Graphic* staff as much as the car's new styling. Their evaluation of the 500's handling was somewhat more reserved

Carroll Shelby's 1969–70 were arguably the most fetching of all his modified Mustangs. GT500 engines breathed fresh air directly from their NACA scoop hoods—no shaker scoops needed.

and references were made to the package's nose-heavy tendency to understeer. Nonetheless, *Sports Car Graphic*'s overall evaluation was on a par with a Siskel and Ebert thumbs up. 1,536 Shelby buyers acted on that (and similar) recommendations in 1969 and drove off the lot in Cobra Jet-powered GT500s.

Interestingly, not all 1969 Shelby Mustangs were completed by the end of regular 1969 production in July of that year. In point of fact, nearly 800 "Shelbys to be" were in the production pipeline when production was shifted to the 1970 model year. In order to freshen those cars for the 1970 sales floor, twin black hood stripes and a plastic chin spoiler were mounted. Each of the cars' VIN numbers were also updated to reflect the new model year. But beyond those few modifications, the 789 1970 Shelby GT350s and 500s built remained unchanged from their original 1969 configurations. Exact 1970 production totals are unclear, it's been estimated that just short of 300 GT500 fastbacks and convertibles were built that year. After them, there was only the darkness of the restrictive 1970s and the horrors of the Maverick Grabber.

The GT500 control cabin was built Mach 1 plush. Better yet, it came factory outfitted with Shelby door and dash panel badges. Note the rim blow steering wheel in this GT.

Leftover 1969 GT500s received a cosmetic re-do and were then sold as 1970 models. Part of that make over was the installation of twin black hood stripes. Flashy 15x7 Mag Star rims were part of the GT500 package in both 1969 and 1970.

The Mach 1 styling package was nearly identical to the stripe and black out treatment found on Boss 351s in 1971. Especially when Boss style hockey stick stripes were ordered.

SEVEN

The 429 Cobra Jet Mustang

Time ran out for Mustang high performance in 1971. Although the newly redesigned coupes, fastbacks, and convertibles that debuted that year had originally been designed to accommodate a wide variety of high horsepower engines—everything from Boss 302s to Boss 429s in fact—by the time final powertrain selections were made, just two muscle car motors made the cut. One of those engines was the Boss 351 small block discussed elsewhere in this work. The other was the last big-block engine to come factory stock in a Ford pony car, the 429 Cobra Jet.

Evolved from the same "385" engine family that had spawned the Boss 429 NASCAR motor, the 429 Cobra Jet (and Super Cobra Jet) engine first rumbled to life under the hoods of 1970 Ford intermediates, mainly because the Fairlane and Montego lines were on a slightly different design change schedule than their Mustang and Cougar little brothers. When the Mustang (and Cougar) grew in size for 1971, its engine box was also enlarged a sufficient amount to accommodate the new and physically larger 385 engine series. Fact of the matter is, the shock towers under the hoods of all 1971 Mustangs had been configured to accept even the big bad Boss 429 engine, so sliding a 429 Cobra Jet into place was no struggle at all.

Like the Boss 429, a 429 Cobra Jet engine was built around a thin wall cast, short-skirted block that featured (in 1971 Mustang trim) beefy four-bolt mains. Unlike the Boss motor, which carried four main caps machined for a quartet of fasteners, a Cobra Jet engine used four bolt caps on just the middle three crank journals. Cobra Jet blocks, identified by the casting numbers "DOVE-A," also lacked the screw-in freeze plugs common to the Boss 429, but they were plenty sturdy, nonetheless. A high nodularity 3.59in stroke crank shaft that had been Brinnel tested for hardness was used to

fill up the empty spaces in the bottom of each block and it was fitted with eight forged rods that had been spot faced for high tensile strength and "football" headed bolts. Cast aluminum, flat top pistons filled each of the engine's cylinder bores and produced a high test-craving 11.3:1 compression ratio.

Base Cobra Jet engines relied on a "juice" cam to oversee timing events and like most of the Hi-Po cams Ford used during the muscle car era, it carried a dual-pattern grind (duration was 282 degree intake, 296 degree exhaust; lift was .506in intake and exhaust) that favored the exhaust side of the internal combustion process. Once in rotation that cam acted on valve train componentry that was carried in what were arguably the best all around big-block heads that Ford ever cast up. Identified by the casting numbers DOOE-R, 429 Cobra Jet (and Super Cobra Jet) heads featured huge, 2.12x2.56in intake ports and equally generous, 1.32x2.24in exhaust passages. Flapjack-sized 2.24in intake and 1.72in exhaust valves were also part of the Cobra Jet package and, like the flow controllers used in the Boss 302 and 351 Cleveland lines, those valves were mounted in poly-angle semi-hemi fashion. Screw-in, non-adjustable rocker studs, stamped steel rockers, pushrod guide plates, and factory-installed rev cups were used to dress out the heads. A racy set of cast aluminum covers were used to button up the rocker area, and they worked in concert with a baffled oil pan to keep internal surfaces clean. Interestingly, Ford chose not to make a crankshaft windage tray part of the 429 program.

Cobra Jet combustibles were brewed up by a 715cfm Ford facsimile of the spread-bore "Rochester" carburetor bolted to a cast iron high-rise dual plane manifold flanged specifically for it. The other half of the internal combustion process was delegated to a pair of cast iron "header style" exhaust manifolds that dumped into a 2¼in dual exhaust system.

Ram Air Cobra Jets used a rubber seal to mate their air cleaners to a plastic plenum that was mounted under the hood. Twin NACA style scoops fed that plenum when the car was in motion. Note the alloy valve covers that Cobra Jet engines carried in 1971.

1971 429 Cobra Jet Road Test Results

Cars magazine
    zero to 30.....4.4sec
    60.....6.5sec
    ¼ mile 14.15sec@102.00mph

As with the 428 Cobra Jet engines that had gone before, 429 Cobra Jets were available with and without Ram Air induction. Just as in 1968, Ram Air cars came equipped with an air cleaner base that sealed to the underside of the hood with the help of a rubber ring. Outside air was provided by a large plastic plenum that ran beneath the hood from the air cleaner opening to a pair of NACA-inspired scoops mounted towards the front of the car. Vacuum controlled diaphragms carried within that duct work allowed fresh air to enter the engine when the throttle was flat on the floor. Non-Ram Air-equipped cars were identified (whether Cobra Jet or Super Cobra Jet) by a "C" as the fifth digit in their serial numbers while a "J" was used to designate ram air ponies. Ford chose not to distinguish between "C" and "J" Cobra Jet engines and rated both a 370hp.

Five more ponies were attributed to the Super Cobra Jet engines Ford offered to Mustang buyers in 1971. Like their slightly less enthusiastic Cobra Jet stable mates, Super Cobra Jet engines were based on beefy "DOVE-A" four bolt blocks, Brinnel tested cranks and large ported "DOOE-R" head castings. The two engines did differ in a number of significant ways, however. Super Cobra Jet engines sported heavy-duty forged aluminum pistons and a hotter, solid lifter camshaft (300 degree exhaust/300 degree intake; .509in exhaust and intake) and an attendant set of adjustable rockers. Super Cobra Jet engines also relied on 780cfm Holley four barrels carried on cast iron intakes that had been machined specifically to accommodate their four equally spaced throttle bores. As in preceding years, the Super Cobra Jet engine was conjured up by ordering either of the two (3.91 or 4.11) "optional axle ratios" in the RPO book. As in 1969 and 1970, a Super Cobra Jet buyer also received an auxiliary oil cooler that was plumbed to the engine via a set of high pressure lines and a special AN fitting-equipped oil filter adapter. Horsepower ratings for the Super Cobra Jet engine remained stable at 375 whether a "Drag Pack" equipped car was outfitted with Ram Air induction or not.

Bright colors like Grabber yellow helped set Cobra Jet ponies off from the crowd in 1971. A body color bumper was standard for Cobra Jet Mach 1s in 1971. 15x7 Magnum 500s were optional. .

Other bits and pieces of the 1971 Cobra Jet/Super Cobra Jet package included a rev limiter, a nodular 9in differential, and a pair of sturdy, 31-spline axles. One final choice the Cobra Jet buyer had to make was his selection of torque reducers. Those set on shifting for themselves could call up one of Ford's tried and true Top Loader four speeds, complete with a "T" handled Hurst shifter. A beefed up C-6 was also a possibility for 1971, and when used to back up a 429, Ford's biggest automatic came filled with a variety of "extra duty" innards.

True to its previous iterations, a 429 Cobra Jet (or Super Cobra Jet) was an engine package rather than a separate car line. And as a result, Ford's biggest wedge motor was orderable in any trim level or Mustang body style that suited a particular buyer. The plushest of the lot was the Mach 1 package, which was changed significantly for the new model year. The 429-powered Mach 1s came standard with the twin-scooped NACA hood, with or without the Ram Air option that made that bonnet functional. A large color-keyed black-out (or argent-hued) panel was also part of the Mach 1 package for 1971. The color-keyed theme was also carried over to the car's front urethane-covered bumper, as well as its hood and fender moldings. A honeycomb grille was used to fill up the radiator opening and, as in 1970, that plastic molding carried a pair of fake fog lights. Moving farther aft, the Mach 1 treatment included color-keyed lower body paint (in either black or argent) and a ribbon of bright body moldings to set those panels off. Model designation was spelled out by a pair of front fender decals, and the cars lineage was also spelled out in a deck mounted tape stripe that ran the width of the car. Final elements of the package included a honeycomb covered tail light panel, a pop open competition style gas cap, and a flat center cap and trim ring combo that dressed up the standard, stamped steel rims. Magnum 500s were optional as were the quartet of simulated mag wheel hub covers. Front and rear spoilers could also be ordered to create a competition atmosphere. High-back bucket seats covered in two-tone vinyl were the central focus of the 1971 Mach interior package. Padded door panels with teak wood accents were also a standard Mach 1 accoutrement. Interestingly, wood grain dash inserts were not part of the deluxe interior option for 1971. A floor-mounted console (which for 1971 carried a clock), a re-designed fold down sport deck, and a tilt steering column were continued as options for the new model year. And the usual array of power assists and comfort options (air conditioning, power steering and brakes, in-dash tach, etc.) were as orderable in 1971 as they had been in preceding years. What was new, was the availability of electric windows, which when checked off on the option sheet resulted in power-operated side and quarter window glass.

Of course, the Cobra Jet engine could also be fitted to a base pony just as easily as it could be for a Mach 1. But based on the Cobra Jets around today, it

429 Cobra Jet intake ports were generously proportioned to say the least. The free-flowing heads carried 2.24in intake and 1.73in exhaust valves. Cobra Jet and Super Cobra Jet heads carried the casting number D00E-R.

seems that more of the motors came clothed in Mach 1 finery than otherwise.

Motor Trend placed a fire engine red Cobra Jet Mach 1 on the cover of its September 1970 issue over the subtitle "Mach 1: Wildest Fastback." *Sports Car Graphic* was one of the first buff books to test a 429-powered Mach 1 (in October of 1970) and they were impressed with their car's ability to hurtle through the quarter mile lights in just 14.6sec despite its nearly two ton weight (3936lb to be exact). At the same time they lamented the fact that the energetic engine's 720lb mass (the heaviest of all Ford engines built in the performance era) was mounted so far to the front of the chassis that "great smoking understeer" was an inevitable result. That being said, the article still concluded on an enthusiastic note which saw the car called an "exciting... gutty, masculine beasty...that no hairy man's man is going to be able to pass...up without a second look." Unfortunately, Ford only offered the 429 Cobra Jet engine package long enough for just short of 1,300 such "manly" types to buy one. By the middle of the 1971 model year, the option had been summarily dropped. Although Ford management tried to keep the revered Cobra Jet name afloat by shifting it to a low-compression version of the 351 Cleveland motor, they accomplished little more than the sullying of the big-block Cobra Jet's reputation. A small block just didn't have the cojones a fire-breathing big block had, no matter what the factory might call it. And so, the 429 Cobra Jet engine passed into Ford high performance history. It was a sad end to a proud name.

Appendices

Appendix A

Boss and Cobra Jet Engine Specifications

Engine	Weight (lb)	Width (in)	Length (in)	Height (in)
302-4V Boss	500	24.5	29	28.5
351 Cleveland	525	25.5	29	29
352, 390, 427, 428	625	27	32	29
429, 460 (except Boss)	720	27	34	29
429 Boss	635	30	34	30

Boss 302

Produced	1969–70
Engine Family	small-block Windsor
Size	4.00inx3.00in (5.0 Liter)
V/C Bolts	eight
Intake	aluminum, dual plane
Carburetor	Holley manual choke, 780cfm
Camshaft	solid lifter 290 degrees/290 degrees (I/E).477in/.477in (I/E)
Engine Code	G
Compression	10.5:1
Torque	290@4300rpm (advertised)
Horsepower	290@5800rpm (advertised)

Quick Reference Casting Numbers

Block	C8FE, C9ZE, DOZE-B. (D1ZE-B service block)
Camshaft	"VED" mark on end
Carburetor	C9ZF-J, DOZF-Z square Holley flange
Crankshaft	forged steel
	1969 C7ZF AND C9ZF cross-drilled
	1970 DOZE-A not cross-drilled
Distributor	C9ZE-F dual point
Exhaust Manifold	C9ZE-9428-A (R), C9ZE-9431-A (L)
Heads	C9ZE-A, C9ZE-C, DOZE-A, DOZE-E
Intake Manifold	C9ZE-D, C9ZE-E, DOZE-E
Connecting Rods	C9ZE-B OR C9ZE-F, forged steel with ⅜in spot-faced rod bolts
Rev Limiter	
1969	no marking or C9ZF-12450-A (6,150rpm)
1970	DOZF-12450-B (6,150rpm)

Features

The Boss 302 used a 302 Windsor block, except the Boss:
1) Used thicker cylinder walls
2) Used four-bolt main caps on two, three, and four.
3) Used screw-in core "freeze" plugs. The crankshafts were forged steel. The 1969 cranks were cross-drilled, the 1970s were not. The 1969 used balancer P/"C9ZZ-6316-B", which measured 6½ O.D. and was 3²¹⁄₃₂in long. The 1970 balancer, P/"DOZZ-6316-A" measured 6½ O.D.. The connecting rods were 289 in length; however, they used ⅜in spot faced bolts.

The cylinder heads were 351C-4V type except:
1) They routed the water flow through the intake, unlike the 351C which did not route water through the intake. The 1969 Boss heads used a 2.23in intake valve while the 1970 used a 2.19in. Both used a 1.71in exhaust valve. The 1969 used valve spring seats P/C9ZZ-6A536-A while the 1970 used P/DOOZ-6A536-H. Both used: Guideplate P/C9ZZ-6A564-A, pushrod P/C9ZZ-6565-A which were hardened and measured 7.595in, heat treated rocker arms P/C9ZZ-6564-A, screw in studs P/C9ZZ-6A527-A, heavy duty 80 PSI oil pump P/C9ZZ-6600-B, and windage tray P/C9ZZ-6687-B. 1969 Bosses used chromed steel valve covers. Early 1970 Boss 302s used slightly different chromed steel valve covers. Later 1970 Boss 302s used finned aluminum valve covers P/"C9ZZ-6582-C." All Bosses used baffled oil pans.

Boss 351

Produced	1971
Engine Family	335 series small block
Size	4.00x3.50in (5.8 LITER)
V/C Bolts	eight
Intake	aluminum
Carburetor	Ford/Autolite 4300-D
Camshaft	mechanical
	290 degrees/290 degrees(I/E).477in/ .477in(I/E)
Engine Code	R
Compression	11.1:1
Torque	380@3400rpm
Horsepower	330@5800rpm

Quick Reference Casting Numbers

Engine Block	D1ZE-A
Camshaft	marked "BY" at the last lobe and journal

Carburetor	marked "D1ZF-FA,GA, D1ZF-ZA"
Crankshaft	marked "4MA", brinell test mark on counterweight
Distributor	marked "D1ZF-DA".
Exhaust Manifold	cast iron D1ZE-9431-DA(R); D1OE-9430-AA(L)
Heads	marked "D1ZE-B"
Intake	D1ZE-F
Rods	marked "D1ZX-AA," forged from 1041-H steel
Rev Limiter	DOZF-12450-B (6,150rpm)

Features

The Boss 351 was produced in 1971. It used a mechanical camshaft with fully adjustable valvetrain which included: screw-in studs, valve spring seats P/"DOOZ-6A536-H", hardened pushrods and guideplates P/"C9ZZ-6A564-A". The Boss used a closed chamber head. The valves sizes for the heads were: intake 2.19in, exhaust 1.71in. The engine used four-bolt main bearing caps at all positions (1-5). The crankshaft was cast from high nodular iron and Brinell tested to insure high rpm stability. The Boss connecting rod was forged, magnafluxed, shot peened, and used 180,000psi rod bolts. The Boss 351 used balancer P/"D1ZZ-6316-B". The Boss used forged pop-up pistons. The exhaust manifold was standard 351C-4V cast iron. The Boss intake was: aluminum dual plane, Holley bolt pattern, although it used an Autolite carb. It used finned aluminum valve covers and steel baffled oil pans.

428 Cobra Jet, 428 Super Cobra Jet

Produced	1968 1/2 TO 1970
Engine Family	FE big block
Size	4.130x3.984in (428ci)
V/C Bolts	five
Intake	cast iron
Carburetor	Holley 735cfm
Camshaft	hydraulic
	270 degrees/290
	degrees(i/e).481in/.490in(I/E)
Engine Code	"Q" non-ram air, "R" ram air
Compression	10.6:1
Torque	1968 445@3400rpm
	1969–70 440@3400rpm
Horsepower	1968 335@5600rpm
	1969–70 335@5200rpm

Engine Block	extra ribs in main bearing area, thicker mains
Camshaft	C6OZ-B (light blue stripe by #3 journal)
Carburetor	1968 C8OF-AB (automatic); C8OF-AA (manual)
	1969 C9AFU (automatic); C9AFM (manual)
	1970 DOZF-AA or AD (manual)
Crankshaft	IU 1968 1/2 to 11/16/68 on number seven counterweight
	IUA used with 692 gram piston counterweight
	(scj)iub counterweight (Cobra Jet)
Distributor	C7OZ-F, C8OF-D, C8OF-J, C8OF-H, DOZF-C, DOZF-G
Exhaust Manifold	C8OE-9428-D (L), C8OE-9431-B (R)
Heads	C8OE-6090-N, 16-hole exhaust bolt holes
Intake	C8OE-G, cast iron
Rods	CS PI rods 13/32in bolts; Super Cobra Jet 427 Lemans rods
Rev Limiter	DOZF-12450-A (5,800rpm) (1970 only)

Features

The block was the same for Cobra Jet and Super Cobra Jet. It was reinforced in the main web area, as were the main caps. The heads were the same from 1968 1/2 to 1970. They were basically 427 low risers with a unique 16 bolt hole exhaust flange. All 428 Cobra

Jet and Super Cobra Jet heads used casting number C8OE-6090-N. They both used rocker arm shaft P/C3AZ-6563-A, stands P/C2AZ-6531-B "Aluminum" or cast iron type P/C3AZ-6531-A, rocker arm stand long bolt P/C1AZ-6A527-A and short bolt P/C1DZ-6A527-C, valve springs used a dampner (1969–70) P/C9OZ-6513-E. The crankshaft was cast from high nodular iron. There are several different crankshafts for the 428CJ/Super Cobra Jet depending on the piston and rod combination. The Super Cobra Jet used balancer P/C8AZ-6316-C which measured 7 1/2 O.D. and was marked C8AE-6316-C. It also used an additional counterweight P/C9ZZ-6359-A. The Cobra Jet used balancer P/C8AZ-6316-B and crank spacer P/B8AZ-6359-A. The rods for the 428 Cobra Jet were 390 type with a 13/32 rod bolt. The 428 Super Cobra Jet used 427 Lemans rods with capscrew rod bolts. The Cobra Jet used cast pistons, the SCJ's were forged. A windage tray was also available for the 428 P/C9ZZ-6687-A. The 1968 428s used oil pan P/C6oz-6675-A and both the 1969–70 Cobra Jet and SCJs used P/C9ZZ-6675-D. The Cobra Jet used stamped steel chrome valve covers, the SCJ's were usually finned aluminum. All ram air 428 engines were equipped with finned aluminum valve covers and other features.

IMPORTANT: Casting numbers are similar for many "FE" big block parts regardless of engine size. For proper identification look for distinguishing factors.

Boss 429

Produced	1969–70
Engine Family	385 series big block
Size	4.360x3.590in (429ci)
V/C Bolts	10 (aluminum or magnesium valve cover)
Intake	aluminum
Carburetor	Holley
Camshaft	hydraulic or mechanical
Engine Code	Z
Compression	10.5:1
Torque	450@3400rpm (advertised)
Horsepower	375@5200rpm (advertised)

Quick Reference Casting Numbers

Block	C9AE-6015-A
Camshaft	S 282I/296E duration; .506I/.506E lift
	T 300I/300E duration; .509I/.509E lift
Carburetor	735 Holley, manual choke, marked C9AF-S (1969), DOOF-S
	(1970), DOZF-G, H, U, T; 715cfm Holley
Crankshaft	forged steel, cross-drilled.
	"S" engine marked C9AE-A.
	"T" engine marked C9AE-C.
Distributor	dual points, dual vacuum advance; marked C9AF-U, C9ZF-D
Exhaust Manifold	C9AE-9430-A (l), C9AE-9431-A or C9AE-9431-179 (r)
Heads	1969 Boss marked C9AE; 1970 marked DOAE
Intake	aluminum, dual-plane marked C9AE-D
Rods	"T" engine marked C9AE-B; "S" engine marked C9AE-A
Rev Limiter	1969 CAZF-12450-A (6,150rpm)
	1970 DOZF-12450-B (6,150rpm)

Features

The first 279 boss Mustangs received "S" NASCAR engines and the remainder "T" engines. The "S" engine used NASCAR rods featuring a 1/2" rod bolt and weighed 1145 grams. It is also .056in shorter than the "T" rod. The "T" rod uses a 3/8in rod bolt and weighs 814 grams. They both used forged pistons and the "T" engines were marked C9AE-6110-B while the "S" engines were marked C9AE-6110-A. Due to the differences in rod weights, they used different forged/cross-drilled crankshafts. They both used dampner P/C9AZ-6A312-A. The blocks were cast: with a high nickel

125

content, used screw in core "freeze" plugs, had 4 bolt main caps at positions 1, 2, 3, 4 and were notched for pushrod clearance. They also had special oil return provisions. They are marked "HP 429" near the water pump mounting flange. Each block had its own way of sealing the heads. The street engines had O-ring grooves in the head while the NASCAR engines had O-ring grooves in the block. All engines used the same: 100 PSI dual entry oil pump M/C9AE-6621-A and a six quart baffled oil pan M/C9AE-6675-A. The heads marked C9AE used a 1.90in exhaust valve P/C9AX-6505-D and an intake valve P/C9AX-6507-B, which measured 2.37in. The DOAE heads used a 1.90in exhaust valve P/C9AZ-6505-J and intake valves P/C9AZ-6507-J which measured 2.28in. They both used the same rocker arms: intake P/C9AZ-6564-C, 1.65 ratio and exhaust P/C9AZ-6564-D, 1.76 ratio. They both used: studs P/C9ZZ-6A527-B, shafts P/C9AZ-6563-A (int. and exh.), stands P/C9AZ-6531-A for both intake and exhaust. The intake was a dual plane aluminum, and used a 735 Holley carburetor. The 1969 Boss used carburetor tray P/C9ZZ-9A600-A and the 1970 models used P/DOZZ-9A9600-A.

429 Cobra Jet, 429 Super Cobra Jet

Produced	1970–71
Engine Family	385 series "Lima"
Size	4.360x3.590in (429ci)
V/C Bolts	eight per cover
Intake	cast iron
Carburetor	Cobra Jet uses Rochester 715cfm
	Super Cobra Jet uses Holley 780cfm
Camshaft	Cobra Jet uses a non-adjustable hydraulic
	Super Cobra Jet mechanical
Engine Code	"C" is non-ram air; "J" is ram air
Compression	11.3:1
Torque	450@3400rpm (both Cobra Jet and
	Super Cobra Jet)
Horsepower	375@5600rpm (Super Cobra Jet)
	370@5400 (Cobra Jet)

Quick Reference Casting Numbers

Engine Block	1970–71 M/DOVE-A, DIVE-A or
	D1OZ-6010-A
Camshaft	Cobra Jet/Boss 429(S) 282I/296E duration;
	.506I/.506E lift
	Super Cobra Jet Boss 429(T) 300I/300E
	duration; .509I/.509E lift
Carburetor	DOOF-A, DOOF-B, DOOF-E, DOOF-F,
	DOOF-N, DOOF-R, D1ZF-YA OR D1ZZ-XA
Crankshaft	high nodular iron "U
	on fourth counterweight
Distributor	DOOF-AA, DOOF-J, DOOF-Y OR
	D1AF-NA (dual-point)
Exhaust Manifold	marked D1ZE-9431-CA or D1ZE-9431-CA1 or
	D1ZE-9431-CA2 (R); DOOE-9430-A (L)
Heads	all years and models marked DOOE-R
Intake	Holley flange-type M/DOOE-C
	Ford spread-bore M/DOOE-D
	Quadrajet flange marked D1AE-BA
Rods	marked DOVE-A; forged steel, spot-faced
Rev Limiter	Cobra Jet DOZF-12450-A (5,800rpm)
	Super Cobra Jet DOZF-12450-B (6,150rpm)

Features

The 429 Cobra Jet/Super Cobra Jet blocks usually use the block casting M/DOVE-A. This block is exceptionally strong and has extra thick main bearing webbing. The 1970 Cobra Jet blocks were all supposed to have 2 bolt mains. The 1971 Cobra Jet and all SCJs were all supposed to have 4 bolt mains at positions 2, 3, and 4. Some blocks have a "CJ" cast in the lifter valley and some have a large "A" cast on the face of the near block. The oil filter, the crank-

shafts were the same on both engines and were Brinnel tested to ensure highrpm durability. A polished area or white paint mark can be found on one of the counterweights. Both engines shared the same connecting rod which was forged steel and used a 3/8in spot faced rod bolt. The Cobra Jet used cast flat top pistons while the SCJ's were forged. All Cobra Jet and Super Cobra Jet engines use the same head cast DOOE-R. The Super Cobra Jet heads were designed to be fully adjustable and used screw in studs P/C9ZZ-6A527-A. The Cobra Jet head was not adjustable and used stud P/DOZZ-6A527-A. Both engines used guideplates P/DOOZ-6A564-A, dual 315 pound valve spring P/DOZZ-6513-A, valve spring seat cups P/DOZZ-6A536-A and heat treated rocker arms P/C9ZZ-6564-A. All intakes are cast iron. Finally, both engines used a high volume oil pump P/C9AZ-6600-A and a baffled oil pan P/DOZZ-6675-D.

Appendix B

428 Cobra Jet Versus Super Cobra Jet Engine Component Part Numbers

Part	428CJ	428SCJ
Connecting Rod	C6AZ-6200-C	C9ZZ-6200-A (427 cap screw)
Crankshaft	C9ZZ-6303-B to	C9ZZ-6303-A (B/12/26/68)
		12/26/68 C9ZZ-6303-D (F/12/26/68)
	C9ZZ-6303-E from 12/26/68	
Piston	piston "A" 680 grams	piston "D" 692 grams
	before 11/13/68	before 12/26/68
	C8OZ-6108-G	C9ZZ-6108-A
	(red size)	(red size)
	C8OZ-6108-H	C9ZZ-6108-B
	(blue size)	(blue size)
	piston "B" 692 grams	piston "E" 712 grams:
	11/13/68 to12/26/68	after 12/26/68
	C9ZZ-6108-G	C9ZZ-6108-N
	(red size)	(red size)
	C9ZZ-6108-H	C9ZZ-6108-R
	(blue size)	(blue size)

piston "C" 712 grams:after 12/26/68C9ZZ-6108-Y (red size)C9ZZ-6108-Z (blue size)

Piston Rings	C6AZ-6148-A	C6AZ-6148-A
Flywheel (MT)	C8OZ-6375-A	C9ZZ-6375-A
Flexplate (AT)	C6AZ-6375-B	C9OZ-6375-B
Balancer	C8AZ-6316-B	C8AZ-6316-C
Crank Spacer	B8AZ-6359-A	C9ZZ-6359-A w/counterweight

Note: Do not interchange 428 Super Cobra Jet parts with each other or with Cobra Jet parts or imbalance will result.

428 Super Cobra Jet Identification

The 428 Super Cobra Jet used no V.I.N. identification number. The only way to determine an Super Cobra Jet car is by the axle code (found on the warranty plate under "axle"). The 428 Super Cobra Jet used either axle code "V" or "W". Axle code "V" is a 3.91 traction lock and "W" is a 4.33 Detroit locker. These were referred to as the Drag Pack options. All 428 Super Cobra Jet engines were also Drag Pack cars. Further, the Super Cobra Jet used finned aluminum valve covers (Cobra Jet was chrome) and came with an oil cooler in front of the radiator. Cars with manual transmissions used staggered shocks. The engine also received some internal modifications, such as: 427 capscrew LeMans rods, forged pistons and different engine balance. The Super Cobra Jet engine was available with or without Ram Air, although most did have the Ram Air option.

428 Ram Air Option

The 428 "Ram Air" included: functional through the hood shaker air intake scoop, competition suspension (staggered shocks with manual transmissions), 3.50 non-locking rear axle (other gears and combinations were optional) and finned aluminum valve covers. The Ram Air option was available with either Cobra Jet or Super Cobra Jet engines.

Appendix C

Boss And Cobra Jet Mustang Production Figures

Model	Year	Production
Boss 302 Mustang	1969	1,628
Boss 302 Mustang	1970	7,013
Total Boss 302 Mustang		8,641
Boss 429 Mustang	1969	857
Boss 429 Mustang	1970	499
Total Boss 429 Mustang		1,356
Boss 351 Mustang	1971	1,806
Total Boss Mustang (1969–71)		11,803
428 CJ/SCJ Mustang	1968 1/2	2,827
428 CJ/SCJ Mustang	1969	13,193
428 CJ/SCJ Mustang	1970	2,671
Total 428 CJ/SCJ Mustang		18,691
429 CJ/SCJ Mustang	1971	1,250

Index

1969 Boss 302
    Chassis, 29–31, 51
    Engine, 31–35, 45, 50
    Race chassis, 38–42
    Race engine, 27, 28, 30, 33
1969 Boss 429
    Engine, 79–87
    Race engine, 78–80, 86
    Race history, 88–90
1969 Cobra Jet 428
    Engine, 102–112
    Mustang chassis, 109
    Racing, 104–105
1970 Boss 302
    Chassis, 70–71
    Engine, 69–70
    Race engine, 56
1970 Boss 429, 91–93
1970 Cobra Jet 428, 114,
1971 Boss 302 parts, 74
1971 Boss 351
    Chassis, 96–97
    Engine, 95–97
    Racing, 97, 101
1971 Cobra Jet 429, 121–123
351 Cleveland, 18, 95
Acid dipping, 18, 39
Big-block Ford intake port comparison, 105
Boss 429 ram air knob, 84
Boss 429 widened shock towers, 81
Boss cylinder heads, 25
Cammer engine, 77–78
Carburetor, In-Line , 25, 42, 56, 57
Cleveland heads, 32
Cougar, 12, 14, 24, 64, 80
Cross Boss, 25, 42
Dearborn, Metuchen Bosses, production differences, 75
Drag Pack, 112
Dual four-barrel intake, 66
Factory Trans-Am Boss 302s, 43

France, Bill, 78
Gurney, Dan, 13, 57
Iacocca, Lee, 9, 11-12, 55, 67
In-Line carburetor, 25, 42, 56, 57
Jones, Parnelli, 13, 37, 57-59, 65
Kar Kraft, 24, 26-27, 29, 38-39, 41-42, 77, 80, 85, 90
Knudsen, Semon "Bunkie," 21, 23, 46, 55, 79, 90, 95
Mach 1 prototype, 108
Mach 1, 108-109, 114, 115, 120, 122
Mini-plenum intake, 56, 58
Moore, Bud, 12-13, 34, 41, 56, 58, 64-65, 67, 97
Muscle car shootout, 1970, 56
NASCAR, 36, 44, 46, 47, 78, 87, 88, 96-97
NHRA Winternationals, 107
Petty, Richard, 77, 88
Production differences between Dearborn, Metuchen Bosses, 75
Ram Air, 28, 45, 63, 84, 96, 103, 117, 122
Ram box intake, 56, 58
Shelby American, 14, 42
Shelby dual four-barrel intake, 66
Shelby GT500, 104, 116-119
Shelby GT500KR , 116, 117
Shelby race Boss 302, 29
Shelby, 10, 12, 14-15, 17, 27, 29, 35, 37, 40, 55, 66, 104, 116-117
Shinoda, Larry, 21-22, 36, 29, 55, 95
Spoiler II, 87-88
Super Cobra Jet, 106, 110, 123
Talladega , 90
Tasca, Bob, 104–105
Titus, Jerry, 11, 13, 14
Trans-Am racing, 1966–68, 11–17
Trans-Am racing, 1969, 35–48
Trans-Am racing, 1970, 55–67
Tunnel Port 302, 14, 16, 18, 105
Winternationals, 105
Yunick, 17, 26, 31, 42, 46-47